HVAC
Control Systems

Second Edition

Workbook

AMERICAN TECHNICAL PUBLISHERS, INC.
HOMEWOOD, ILLINOIS 60430-4600

Ronnie J. Auvil

HVAC Control Systems Workbook contains procedures commonly practiced in industry and the trade. Specific procedures vary with each task and must be performed by a qualified person. For maximum safety, always refer to specific manufacturer recommendations, insurance regulations, specific job site and plant procedures, applicable federal, state, and local regulations, and any authority having jurisdiction. The material contained is intended to be an educational resource for the user. American Technical Publishers, Inc. assumes no responsibility or liability in connection with this material or its use by any individual or organization.

© 2007 by American Technical Publishers, Inc.
All rights reserved

2 3 4 5 6 7 8 9 – 07 – 9 8 7 6 5 4 3 2

Printed in the United States of America

ISBN 978-0-8269-0758-5

 This book is printed on 30% recycled paper.

Workbook
HVAC Control Systems

Contents

1 HVAC Fundamentals

Review Questions	1
Activities	5

2 Commercial HVAC Systems

Review Questions	9
Activities	15

3 HVAC System Energy Sources

Review Questions	19
Activities	25

4 Control Principles

Review Questions	29
Activities	33

5 Control Systems

Review Questions	37
Activities	41

6 Air Compressor Stations

Review Questions	45
Activities	49

7 Pneumatic Actuators, Dampers, and Valves

Review Questions	51
Activities	55

8 Pneumatic Thermostats, Humidistats, and Pressure Switches

Review Questions	59
Activities	65

9 Pneumatic Transmitters

Review Questions	69
Activities	73

10   Pneumatic Receiver Controllers

Review Questions **79**
Activities **83**

11   Pneumatic Auxiliary Devices

Review Questions **87**
Activities **91**

12   Pneumatic Control System Applications

Review Questions **97**
Activities **103**

13   Electrical Control Systems

Review Questions **109**
Activities **113**

14   Electronic Control Systems

Review Questions **117**
Activities **121**

15   Building Automation Systems and Controllers

Review Questions **125**
Activities **129**

16   Operator Interfaces

Review Questions **133**
Activities **137**

17   Building Automation System Inputs and Outputs

Review Questions **143**
Activities **147**

18   Building Automation System Installation, Wiring, and Testing

Review Questions **155**
Activities **159**

19   Networking and Web-Based Control

Review Questions **165**
Activities **169**

20   Direct Digital Control Strategies

Review Questions **171**
Activities **175**

21 Supervisory Control Strategies

Review Questions **179**
Activities **183**

22 Building Automation System Retrofit of Existing Systems

Review Questions **187**
Activities **191**

23 Building System Management

Review Questions **193**
Activities **195**

24 Utilities and Surveys

Review Questions **199**
Activities **203**

25 Building Automation System Troubleshooting

Review Questions **205**
Activities **209**

26 Advanced HVAC Control Technologies and Interoperability

Review Questions **213**
Activities **217**

Name: _____ Date: _____

Control System Principles

_____ 1. ___ is the condition that occurs when people cannot sense a difference between themselves and the surrounding air.

T   F   2. Temperature is the measurement of the heat intensity of a substance.

_____ 3. ___ is the process that occurs when a liquid changes to a vapor by absorbing heat.
   A. Ventilation
   B. Conduction
   C. Radiation
   D. Evaporation

_____ 4. Comfort is usually attained at normal cooling and heating temperatures with a humidity level of about ___.
   A. 35%
   B. 50%
   C. 65%
   D. 70%

_____ 5. ___ is the temperature of air below which moisture begins to condense from the air.

_____ 6. Filtration is the process of removing ___ from air that circulates through an air distribution system.

_____ 7. Air in a building must be circulated ___ to provide maximum comfort.
   A. at a low velocity
   B. at a high velocity
   C. intermittently
   D. continuously

_____ 8. ___ is the process of introducing fresh air into a building.

T   F   9. Makeup air is air that is used to replace air that is lost to psychrometrics.

_____ 10. ___ is heat identified by a change of state and no temperature change.

_____ 11. ___ is heat transfer that occurs when molecules in a material are heated and the heat is passed from molecule to molecule through the material.

_____ **12.** A British thermal unit is the amount of heat energy required to raise the temperature of ___ of water ___.

     A. 1 lb; 1°F
     B. 1 lb; 100°F
     C. 1000 lb; to desired temperature
     D. 1 ton; in 24 hr

_____ **13.** ___ is heat transfer that occurs when currents circulate between warm and cool regions of a fluid.

_____ **14.** ___ is heat transfer in the form of radiant energy (electromagnetic waves).

_____ **15.** ___ is the elements that make up atmospheric air with moisture and particles removed.

     A. Standard air
     B. Dry air
     C. Free air
     D. Exhaust air

_____ **16.** Dry bulb temperature (db) is the temperature of the air without reference to the ___.

T   F   **17.** Relative humidity is the amount of moisture in the air compared to the amount of moisture that it could hold if it were saturated (full of water).

_____ **18.** Wet bulb temperature is measured using a ___.

_____ **19.** ___ is the force created by a substance per unit of area.

_____ **20.** Absolute pressure is pressure above ___.

     A. atmospheric pressure
     B. gauge pressure
     C. a water column
     D. a perfect vacuum

Calculating Pressure

_____ **1.** Box A pressure = ___ psi.

_____ **2.** Box B pressure = ___ psi.

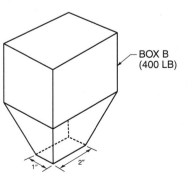

BOX A
(25 LB)

BOX B
(400 LB)

Fahrenheit to Celsius Temperature Conversion

_____ **1.** A temperature of 95°F = ___°C.

_____ **2.** A temperature of 140°F = ___°C.

_____ **3.** A temperature of 55°F = ___°C.

_____ **4.** A temperature of 74°F = ___°C.

_____ **5.** A temperature of 68°F = ___°C.

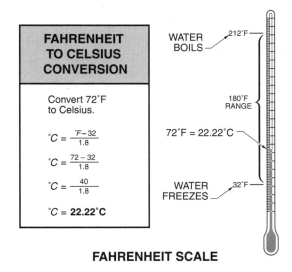

Celsius to Fahrenheit Temperature Conversion

_____ **1.** A temperature of 20°C = ___°F.

_____ **2.** A temperature of 65°C = ___°F.

_____ **3.** A temperature of 10°C = ___°F.

_____ **4.** A temperature of 27°C = ___°F.

_____ **5.** A temperature of 40°C = ___°F.

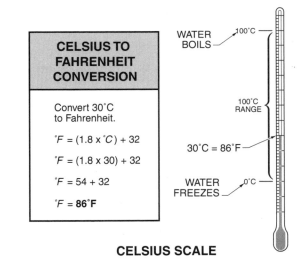

Circulation

_____	1.	filter	_____	7.	heating/cooling coil
_____	2.	circulating air	_____	8.	return air
_____	3.	supply air	_____	9.	supply air fan
_____	4.	return air grill	_____	10.	supply duct
_____	5.	outside air in	_____	11.	supply air register
_____	6.	return duct			

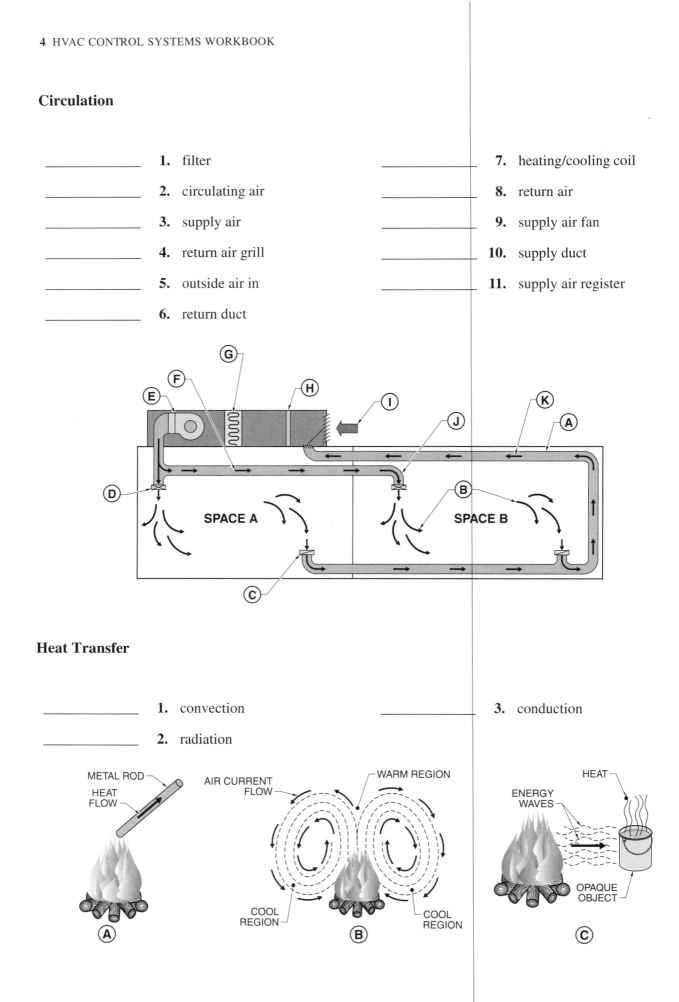

Heat Transfer

_____	1.	convection	_____	3.	conduction
_____	2.	radiation			

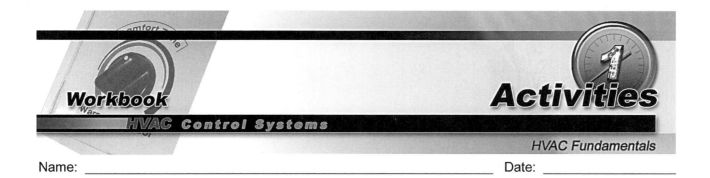

Name: _____ Date: _____

Activity 1-1. Comfort

A complaint is received that a room is too hot. It is 7 AM in the summer and work has just started. The room conditions are checked with an electronic hand-held temperature and humidity meter. The temperature is 85°F and the humidity is 55% rh.

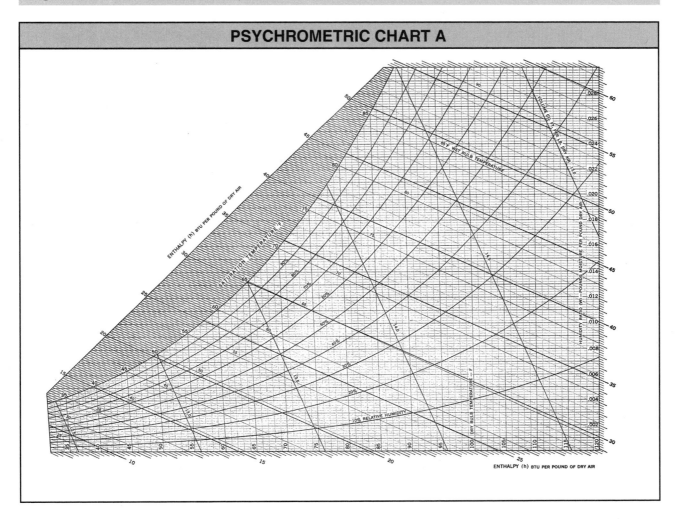

PSYCHROMETRIC CHART A

1. Plot this point on psychrometric chart A.

2. _____ Is this point outside the normal comfort zone?

After checking the building automation system workstation computer, it was found that the time schedule was incorrect, keeping the cooling unit OFF. The cooling unit is switched ON, and after 10 min the discharge air conditions are checked in the room. The temperature is 72°F and the humidity is 45% rh.

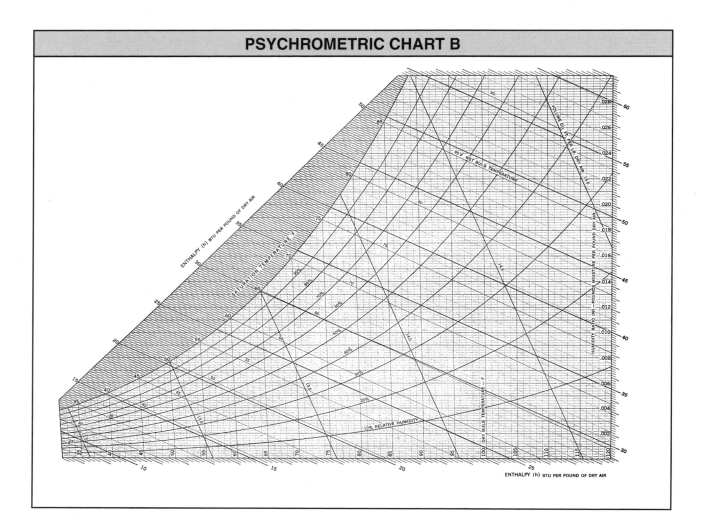

PSYCHROMETRIC CHART B

3. Plot this point on psychrometric chart B.

4. _____ Is this point outside the normal comfort zone?

To monitor systems operation, a data trend is set up to monitor the room temperature and humidity every 15 min. The trend begins at 8 AM and ends at 10 AM. At 10 AM the data is checked.

ROOM TEMPERATURE AND HUMIDITY DATA TREND		
TIME	TEMPERATURE*	HUMIDITY†
8:15 AM	85	55
8:30 AM	83	55
8:45 AM	81	52
9:00 AM	80	52
9:15 AM	78	48
9:30 AM	76	45
9:45 AM	75	42
10:00 AM	74	41

* in °F

† in % rh

PSYCHROMETRIC CHART C

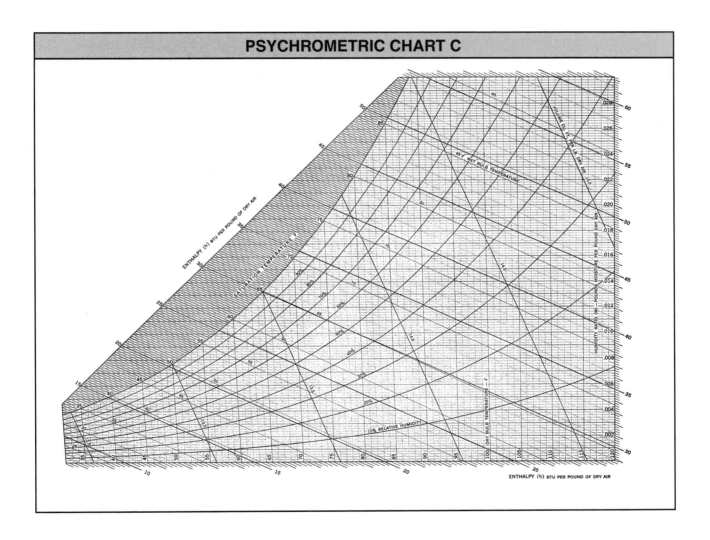

5. Plot the points as closely as possible on psychrometric chart C, then connect them in a straight line.

6. _____ Are the conditions at 10 AM outside the normal comfort zone?

Workbook

HVAC Control Systems

Review Questions

Commercial HVAC Systems

Name: _____ Date: _____

Control System Principles

T   F   **1.** A commercial HVAC system may contain a heating system, ventilation system, cooling system, humidification system, dehumidification system, and/or air filtration system.

_____ **2.** A(n) ___ is a device that transfers heat from one substance to another substance without allowing the substances to mix.

_____ **3.** A(n) ___ is a device that uses an electric resistance heating element embedded in a ceiling panel.

        A. air economizer
        B. rooftop package
        C. radiant heat panel
        D. baseboard heater

_____ **4.** A heat pump is a direct expansion refrigeration system that contains devices and controls that ___ the flow of refrigerant.

_____ **5.** ___ is the combination of return air and outside air.

_____ **6.** Common percentages of outside air used for ventilation are ___.

        A. 5%, 10%, and 30%
        B. 10%, 25%, and 50%
        C. 25%, 50%, and 75%
        D. 30%, 40%, and 50%

_____ **7.** An outside air dry bulb temperature of ___ is common for outside air economizer use.

        A. 33°F to 45°F
        B. 35°F to 55°F
        C. 45°F to 65°F
        D. 60°F to 70°F

_____ **8.** ___ is cooling produced by the vaporization of refrigerant in a closed system.

T   F   **9.** Passive dehumidification is the process in which air contacts a chemical substance (desiccant) that adsorbs moisture.

_____ **10.** A(n) ___ is a device consisting of a fan, ductwork, filters, dampers, heating coils, cooling coils, humidifiers, dehumidifiers, sensors, and controls to condition and distribute air throughout a building.

_____ **11.** A(n) ___ is a device with rotating blades or vanes that move air.

    A. damper
    B. economizer
    C. desiccant wheel
    D. fan

_____ **12.** A filter is a porous material that removes ___ from a moving fluid.

_____ **13.** A(n) ___ is a device that cleans air by passing the air through electrically charged plates and collector cells.

_____ **14.** A damper is an adjustable metal blade or set of blades used to control the ___ of air.

    A. dynamic pressure
    B. static pressure
    C. flow
    D. temperature

T   F   **15.** A baseboard heater is a heat exchanger that transfers heat between a medium at two different temperatures.

_____ **16.** ___ air handling units control building space temperature by changing the temperature of air, not the volume.

_____ **17.** A(n) ___ is an air handling unit that delivers air at a constant 55°F temperature to building spaces.

_____ **18.** Many ___ air handling units are used for cooling only, and any heat required is provided using return air.

    A. multizone
    B. terminal reheat
    C. induction
    D. variable air volume

T   F   **19.** An electric motor variable-frequency drive is an electronic device that controls the direction, speed, and torque of an electric motor.

_____ **20.** A(n) ___ is a device that controls the air flow to a building space, matching the building space requirements for comfort.

Hot Water Heating Systems

_____ 1. circulating pump

_____ 2. hot water flow

_____ 3. hot water to other building spaces

_____ 4. piping system

_____ 5. hot water boiler

_____ 6. makeup water supply line

_____ 7. hot water from other building spaces

_____ 8. heating unit

_____ 9. backflow preventer

_____ 10. expansion tank

_____ 11. branch line

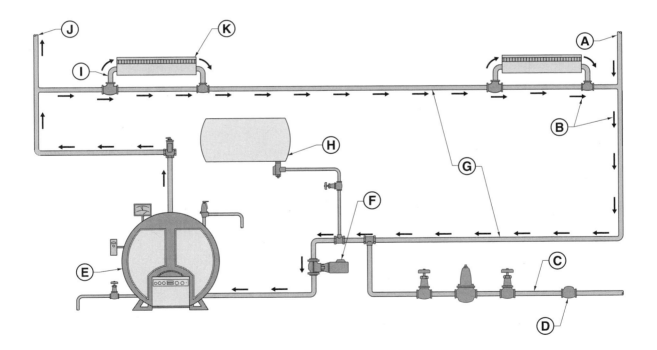

Rooftop Packaged Units

_____ **1.** condenser

_____ **2.** natural gas supply

_____ **3.** filter section

_____ **4.** exhaust hood

_____ **5.** compressor

_____ **6.** return air

_____ **7.** heat exchanger

_____ **8.** cooling section

_____ **9.** supply air fan

_____ **10.** evaporator coil

_____ **11.** supply plenum

_____ **12.** return air fan

_____ **13.** supply air to building space

_____ **14.** outside air louvers

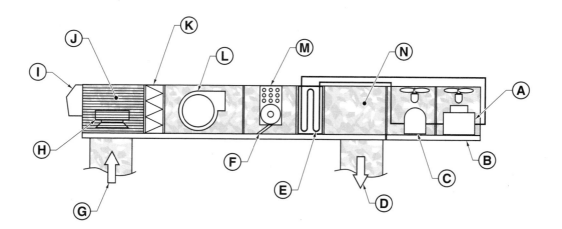

Air Handling Unit

_____	**1.** mixed air plenum
_____	**2.** return air
_____	**3.** humidifier
_____	**4.** damper actuators
_____	**5.** ductwork
_____	**6.** outside air
_____	**7.** fan
_____	**8.** return air damper (NO)
_____	**9.** exhaust air

_____ **10.** heating coils

_____ **11.** mixed air

_____ **12.** outside air damper

_____ **13.** filter

_____ **14.** exhaust air damper

_____ **15.** cooling coils

_____ **16.** temperature controller in building space

_____ **17.** supply air

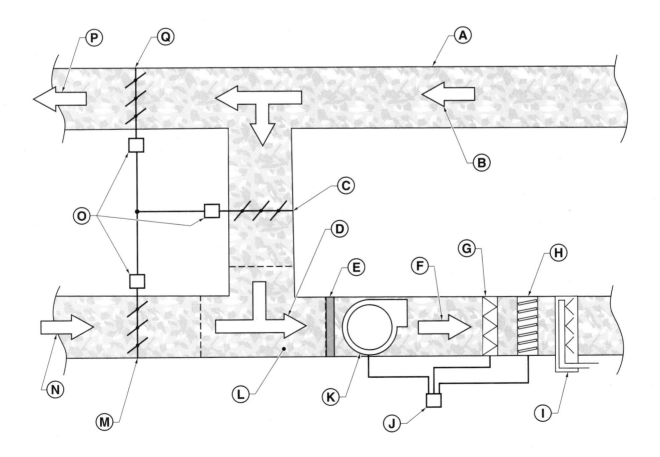

Direct Expansion Cooling

_____ **1.** cool air flow

_____ **2.** warm vapor refrigerant

_____ **3.** evaporator

_____ **4.** evaporator fan

_____ **5.** vaporizing refrigerant

_____ **6.** warm air flow

_____ **7.** cool liquid refrigerant

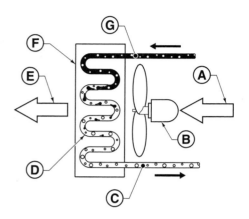

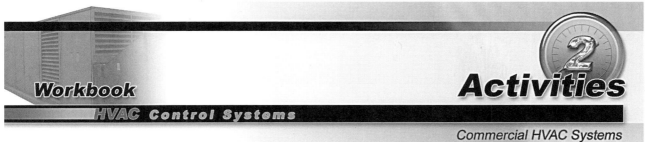

Name: _____ Date: _____

Activity 2-1. Maintenance Checklist

Use the maintenance checklist to answer the questions.

MAINTENANCE CHECKLIST — COOLING TOWERS														
ONCE EVERY MONTH Date:	JAN 1/4	FEB 2/10	MAR	APR	MAY	JUN	JUL	AUG	SEPT	OCT	NOV	DEC		
1. Check fan and motor bearings and lubricate if necessary.	RA	BW												
2. Check tightness and adjustment of thrust collars on sleeve bearing units and locking collars on ball bearing units.	RA	BW												
3. Check belt tension and adjust if necessary.	RA	BW												
4. Clean strainer (if atmosphere is extremely dirty, it may be necessary to clean strainer weekly).	RA	BW												
5. Check for biological growth in sump. Consult water treatment specialist if such growth is not under control.	RA	BW												
6. Clean and flush sump.	RA	BW												
7. Check spray distribution system. Check and re-orientate nozzles, if necessary. On evaporative condensers and industrial fluid coolers with trough type distribution systems, adjust and flush out troughs if necessary.	RA	BW												
8. Check operating water level in the pan and adjust float valve if required.	RA	BW												
9. Check bleed-off rate and adjust if necessary.	RA	BW												
10. Check fans and air inlet screens and remove any dirt or debris.	RA	BW												
ONCE EVERY YEAR	Inspect and clean protective finish inside and out. Look particularly for any signs of spot corrosion. Clean and refinish any damaged protective coating.													
Before undertaking start-up procedures or performing inspection or maintenance of equipment, make certain the power has been disconnected. Refer to appropriate operating and maintenance manuals and comply with all caution label instructions.														

1. _____ The preventive maintenance checklist procedure is for a(n) ___.

2. _____ Where would the equipment be commonly located?

3. _____ What type of mechanical cooling equipment uses this device?

4. _____ When was preventive maintenance last performed?

5. _____ If it is currently June, how many months have been skipped?

6. _____ If this procedure takes 2 hr per month at a billed rate of $75 per hour, how much money should be budgeted per year for the preventive maintenance labor on this unit?

Activity 2-2. Commercial HVAC Systems

Your firm has taken over the operations and maintenance contract on a commercial facility. In preparation for the beginning of the contract, the type of HVAC equipment must be determined and basic questions answered. Upon arriving at the facility, it is discovered that there are four air handling units in the fan room. Each has an identical diagram. Use the air handling unit drawing to answer the questions.

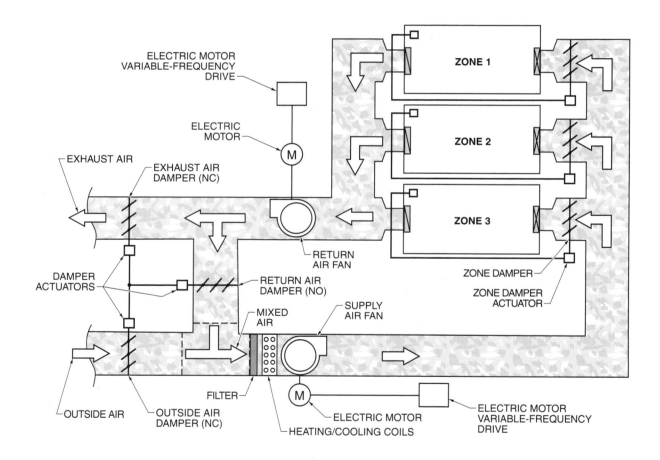

1. _____ What type of air handling unit is it?

2. _____ Is the air handling unit a 100% outside air system or a mixed air system?

3. _____ Is economizer cooling available with the unit?

4. _____ If indoor air quality is a concern, what may be introduced to reduce IAQ problems?

5. _____ What is a common duct static pressure?

6. _____ The unit provides air for ___ (number) zones.

7. _____ If each air handling unit is identical, what is the total number of zones?

8. _____ If a trap is not present on the heating coil, what type of heat is being used?

9. _____ What device removes particulate matter from the air stream?

10. _____ Preventive maintenance has been estimated at ½ hr per month per terminal box and 1 hr for each air handling unit. How many total hours of preventive maintenance per month are required for all of the air handling units and zone terminal boxes?

Name: _____ Date: _____

Control System Principles

_____ **1.** In most HVAC applications, it is more expensive to heat a building space with ___ than with other energy sources.

    A. electricity
    B. natural gas
    C. fuel oil
    D. solar energy

T   F **2.** An electric heating element is a device that consists of pipes that become hot when the steam or hot water valve is energized.

_____ **3.** ___ is commonly used as an energy source for heating commercial buildings because it is plentiful and relatively inexpensive.

_____ **4.** Combustion is the chemical reaction that occurs when ___ reacts with the hydrogen (H) and carbon (C) present in a fuel at ignition temperature.

_____ **5.** ___ is the intensity of heat required to start a chemical reaction.

    A. Btu
    B. Emergency heat
    C. Viscosity
    D. Ignition temperature

_____ **6.** Viscosity is the ability of a liquid to ___.

_____ **7.** ___ is the number of British thermal units (Btu) per pound or gallon of fuel.

    A. Temperature
    B. Heating value
    C. CFC
    D. none of the above

T   F **8.** The four grades of fuel oil used in boilers are No. 2 fuel oil, No. 4 fuel oil, No. 5 fuel oil, and No. 6 fuel oil.

_____ **9.** The fuel oil temperature required depends on the type of ___ and whether a straight distillate fuel oil or a blend of fuel oils is used.

_____ **10.** ___ is the heat created by the visible (light) and invisible (infrared) energy rays of the sun.

_____ **11.** The amount of energy received at the surface of the Earth can exceed ___ per sq ft of surface, depending on the angle of the sun's rays and the position of the solar collector.

    A. 100 Btu/hr
    B. 200 Btu/hr
    C. 300 Btu/hr
    D. 400 Btu/hr

_____ **12.** ___ is heat provided if the outside air temperature drops below a set temperature or if a heat pump fails.

_____ **13.** An air conditioning system is a system that produces a(n) ___ effect and distributes the cool air or water to building spaces.

    A. refrigeration
    B. heating
    C. open flow
    D. heat pump

T   F **14.** The basic energy source used to cool a building is outside air.

_____ **15.** A(n) ___ is a mechanical device that compresses refrigerant or other fluid.

_____ **16.** A(n) ___ is a system that uses a liquid (normally water) to cool building spaces.

_____ **17.** A(n) ___ is a nonmechanical refrigeration system that uses a fluid with the ability to absorb a vapor when it is cool and release a vapor when heated.

_____ **18.** A(n) ___ and ___ are required in absorption refrigeration systems.

    A. condenser; expansion device
    B. generator; compressor
    C. evaporator; compressor
    D. refrigerant; absorbent

_____ **19.** Steam and hot water cooling systems normally do not use ___, avoiding the problem of refrigerant replacement.

_____ **20.** The mechanical equipment of a ___ system can be used to transfer heat from the air inside a building to the air outside a building, producing a cooling effect.

Refrigeration System

_____ 1. condenser   _____ 5. hot gas discharge line

_____ 2. expansion device   _____ 6. compressor/motor

_____ 3. suction line   _____ 7. liquid line

_____ 4. evaporator

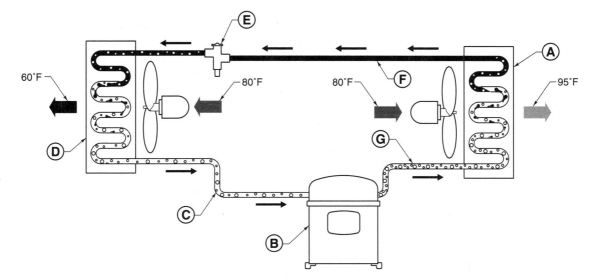

Heat Pump Cooling Cycle

_____ 1. hot outside air

_____ 2. cool air to building space

_____ 3. reversing valve

_____ 4. compressor

_____ 5. warm air from building space

_____ 6. outdoor unit

_____ 7. spool

_____ 8. expansion device

_____ 9. warm outside air

_____ 10. indoor unit

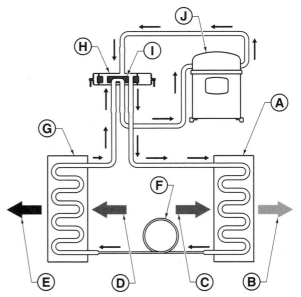

Fuel Oil System

_____ 1. fuel oil relief valve

_____ 2. fuel oil from tank

_____ 3. fuel oil strainer

_____ 4. main fuel oil solenoid valve

_____ 5. fuel oil burner pressure gauge

_____ 6. nozzle air pressure gauge

_____ 7. fuel oil pump

_____ 8. burner nozzle

_____ 9. atomizing air

_____ 10. fuel oil thermometer

_____ 11. fuel oil pressure gauge

_____ 12. fuel oil pressure regulator

_____ 13. check valve

_____ 14. fuel oil returned to tank

_____ 15. shutoff valve

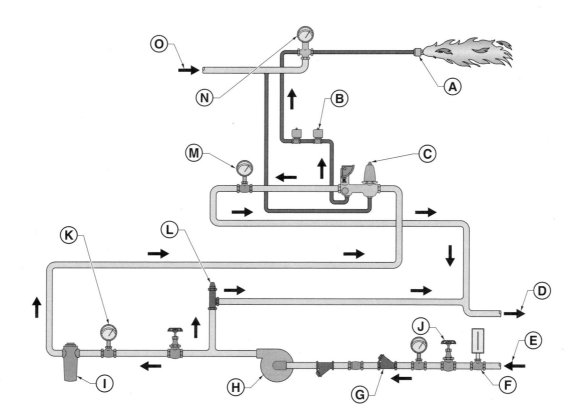

Liquid Chiller

_____	**1.** cooling tower
_____	**2.** warm water out
_____	**3.** liquid chiller
_____	**4.** expansion device
_____	**5.** louver
_____	**6.** air outlet
_____	**7.** cool water collection
_____	**8.** compressor
_____	**9.** hot water inlet

_____ **10.** condenser (tube-in-shell) heat exchanger

_____ **11.** condenser water pump

_____ **12.** cold water used for cooling

_____ **13.** cool water

_____ **14.** evaporator (tube-in-shell) heat exchanger

_____ **15.** refrigerant flow

_____ **16.** air inlet

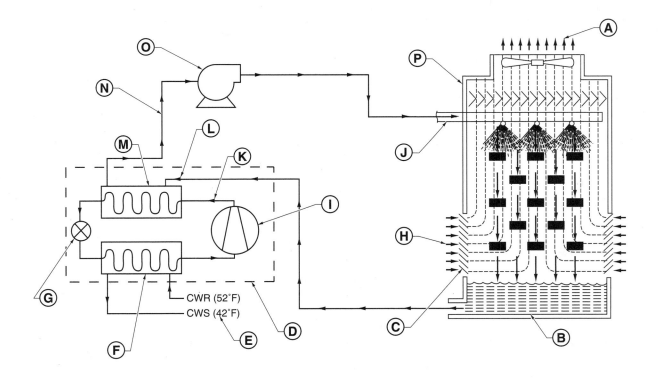

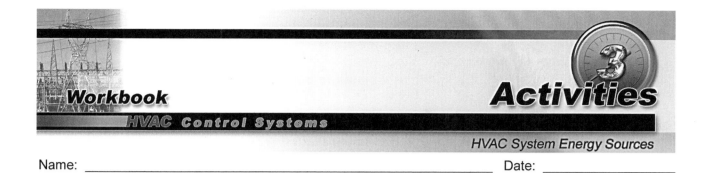

Name: _____ Date: _____

Activity 3-1. Boiler Nameplate

Use the boiler nameplate to answer the questions.

(S)	**STEAMWORKS BOILER CO.** Hilliard, Ohio			**NB**
MODEL NO. 10-13-89 APW		SERIAL NO. 4662-678		
MAXIMUM ALLOWABLE WORKING PRESSURE 250 PSI		HEATING SURFACE 5657 SQ FT		
INPUT	19,125,090 BTU/HR		NAT GAS	
	130 GAL./HR		#2 OIL	
DATE BUILT MAR. 1991		NATIONAL BOARD NO. 2156		

1. _____ The boiler manufacturer is ___.

2. _____ The boiler model number is ___.

3. _____ The boiler serial number is ___.

4. _____ The boiler burner rating is ___ Btu/hr.

5. _____ The listed heating surface area of the boiler is ___ sq ft.

6. _____ The boiler is designed to use ___ gas fuel.

7. _____ What weight oil can be used?

8. _____ How many gallons of oil are used per hour?

Activity 3-2. Heat Pump Selection

Work is required in a new wing of a building that contains heat pumps. The first heat pump nameplate is listed as W-CDD-2-015-F-Z. Use the model nomenclature to answer the questions.

Model Nomenclature

Note: For illustration purposes only. Not all options available with all models.
Please consult Water Source Heat Pump Representative for specific availability.

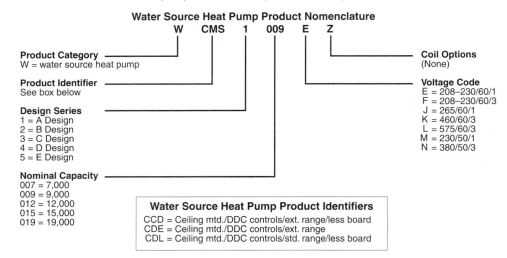

1. _____ Is the heat pump a water source heat pump?

2. _____ Does the heat pump use DDC controls?

3. _____ The heat pump nominal capacity is ___ Btu/hr.

4. _____ If there are 12,000 Btu/hr/ton, how many tons of cooling can be achieved by the heat pump?

5. _____ If the voltage applied is 208 VAC–230 VAC, 60 Hz, 1φ, does this agree with the listing on the nameplate?

Activity 3-3. Absorption Cooling Selection

Absorption chiller information must be determined from the manufacturer's catalog. The absorber model number is T1B-ST-9E2-28-A-S. Use the absorption chiller nomenclature to answer the questions. Note: More data would be included with the absorber.

Model Nomenclature

Note: The model number denotes the following characteristics of the unit:

TIB   ST   8E1   46   A   S

Unit Type

Heat Source
ST = Steam
HW = Hot Water

Unit Size

Special
Special Tubes
Contract Job

Design Level

Voltage Code
17 = 208/60/3
28 = 230/60/3
46 = 460/60/3
50 = 380/50/3
58 = 575/60/3

1. _____ The manufacturer type designation for the absorber is ___.

2. _____ The heat source is ___.

3. _____ The unit size is ___.

4. _____ The voltage used is ___.

5. _____ The design level designation is ___.

6. _____ Does the absorber have special tubes, or is it a contract job?

Name: _____ Date: _____

Control System Principles

_____ **1.** A(n) ___ is an arrangement of a sensor, controller, and controlled device to maintain a specific controlled variable value in a building space, pipe, or duct.

_____ **2.** A sensor is a device that measures a controlled variable such as temperature, pressure, or humidity and sends a signal to a(n) ___.

    A. controlled device
    B. anticipator
    C. actuator (valve or damper)
    D. controller

T   F **3.** A controller is a device that receives a signal from a sensor, compares it to a setpoint value, and sends an appropriate output signal to a controlled device.

_____ **4.** A(n) ___ is the object that regulates the flow of fluid in a system to provide the heating, air conditioning, or ventilation effect.

_____ **5.** A(n) ___ is fluid that flows through controlled devices to produce a heating or cooling effect.

_____ **6.** Toxins and pollutants may build up to unacceptable levels, causing health problems to occupants, if enough ___ is not circulated in a building space.

    A. aerosol
    B. desiccant
    C. fresh air
    D. return air

_____ **7.** Return air temperature control is more accurate than building space temperature control because the return air is at a temperature which is a(n) ___ of all the temperatures in the building space.

T   F **8.** In differential pressure control, a specific pressure in the system is maintained based on a difference in pressure between two points in the system.

_____ **9.** ___ is the desired value to be maintained by a system.

    A. Control point
    B. Setpoint
    C. Offset
    D. none of the above

_____ 10. A(n) ___ is the actual value that a control system experiences at any given time.

     A. control point
     B. setpoint
     C. offset
     D. controlled variable

_____ 11. ___ is the difference between a control point and a setpoint.

_____ 12. Feedback is the ___ of the results of a controller action by a sensor or switch.

T   F   13. Closed loop control is control in which feedback occurs between the controller, sensor, and controlled device.

T   F   14. Open loop control is control in which no feedback occurs between the controller, sensor, and controlled device.

_____ 15. ___ is control in which a controller produces only a 0% or 100% output signal.

     A. Closed loop control
     B. Anticipator control
     C. ON/OFF (digital) control
     D. Overshooting control

_____ 16. A(n) ___ is a device that turns heating or cooling equipment ON or OFF before it normally would.

_____ 17. ___ is control in which the controlled device is positioned in direct response to the amount of offset in the system.

_____ 18. HVAC electrical controls often use ___.

     A. 6 VAC
     B. 24 VAC
     C. 115 VAC
     D. 230 VAC

_____ 19. Once adjusted, a control system should operate ___.

_____ 20. With the widespread use of automated control systems, much tighter accuracies of ___ are possible.

Volume Control

_____ 1. linkage

_____ 2. fan housing

_____ 3. actuator

_____ 4. inlet shroud vane

_____ 5. inlet shroud vane crank arm

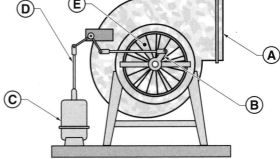

Economizer Cooling

_____ **1.** filter

_____ **2.** open outside air dampers

_____ **3.** closed return air dampers

_____ **4.** supply fan

_____ **5.** open exhaust air dampers

_____ **6.** chilled water valve

_____ **7.** cooling coil

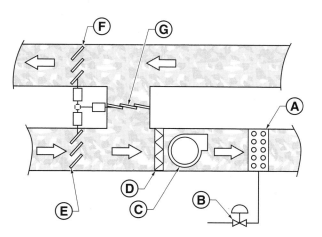

Open Loop Control

_____ **1.** chilled water pump

_____ **2.** chiller evaporator (heat exchanger)

_____ **3.** pump contactor

_____ **4.** outside air thermostat

_____ **5.** chilled water supply

_____ **6.** chiller compressor

_____ **7.** chilled water return

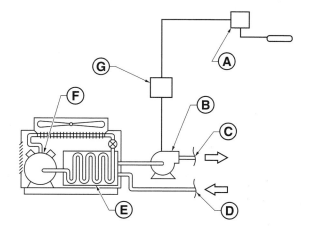

Offset

_____ **1.** control point

_____ **2.** offset

_____ **3.** setpoint

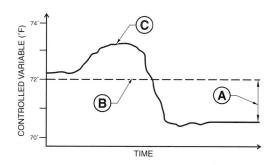

Dehumidification

_____ 1. chilled water valve

_____ 2. cool, dry air

_____ 3. cooling coil

_____ 4. drain pan and drain line to remove condensate

_____ 5. hot, humid air

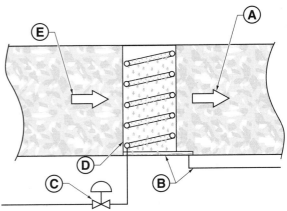

Building Space Temperature Control

_____ 1. filter

_____ 2. supply air duct

_____ 3. exhaust air damper

_____ 4. heating coil

_____ 5. building space

_____ 6. return air duct

_____ 7. outside air damper

_____ 8. humidifier

_____ 9. return fan

_____ 10. cooling coil

_____ 11. room controller or thermostat

_____ 12. supply fan

_____ 13. return air damper

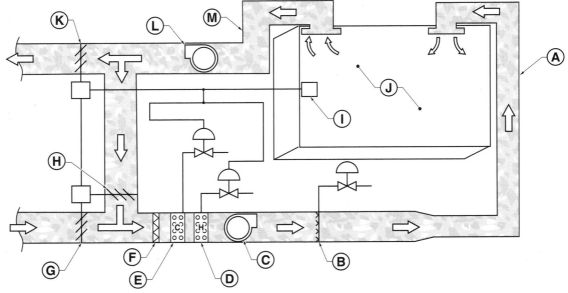

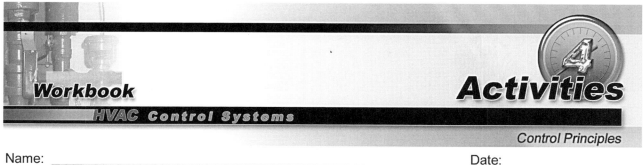

Control Principles

Name: _____ Date: _____

Activity 4-1. Control Principles

The control system of a new building has not been wired.

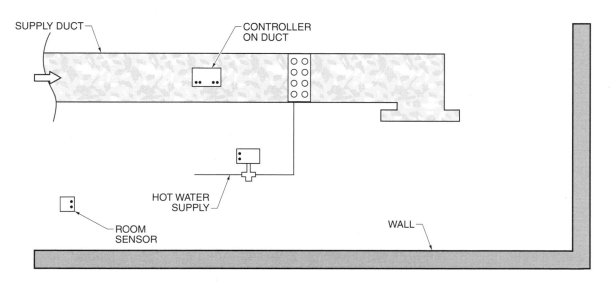

1. Complete the drawing of the room temperature control system by connecting the components to make the control system function.

2. Complete the drawing to show whether the temperature control circuit is open or closed loop control.

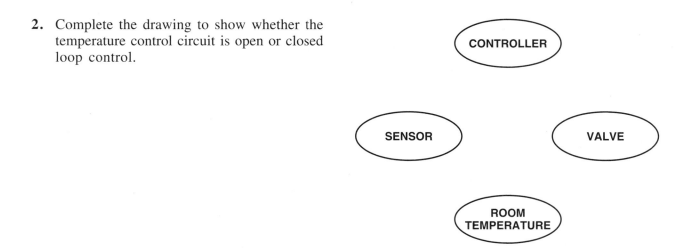

After connecting the devices, the HVAC unit begins to operate. After a few hours, the occupants complain that the temperature is erratic. The room temperature is checked every 5 min to determine system operation. The room temperature is checked with an accurate electronic temperature sensor.

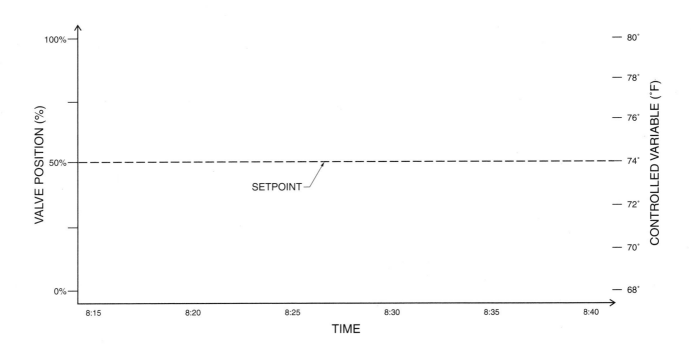

3. Enter the offset value for each temperature measurement.

TEMPERATURE READINGS			
TIME	TEMPERATURE*	SETPOINT*	OFFSET*
8:15 AM	78	74	
8:20 AM	75	74	
8:25 AM	72	74	
8:30 AM	70	74	
8:35 AM	68	74	
8:40 AM	71	74	

* in °F

4. Graph the offset on the chart.

5. _____ Is this level of control acceptable?

After making some controller adjustments, the control appears better. The room temperature is checked again every 5 min to determine system operation.

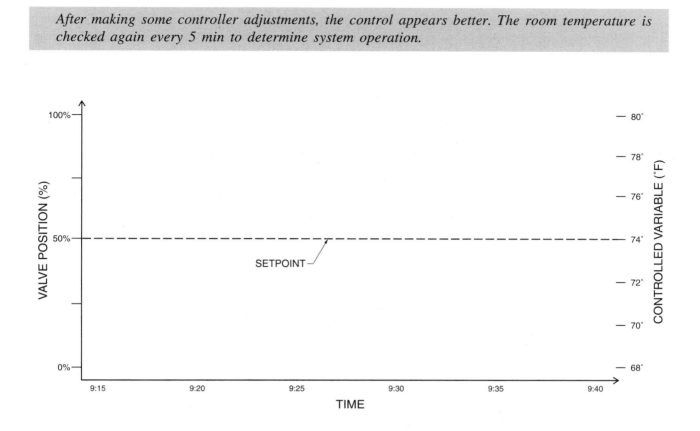

6. Enter the offset value for each temperature measurement.

TEMPERATURE READINGS			
TIME	TEMPERATURE*	SETPOINT*	OFFSET*
9:15 AM	75	74	
9:20 AM	74.5	74	
9:25 AM	74	74	
9:30 AM	75	74	
9:35 AM	75.5	74	
9:40 AM	75	74	

* in °F

7. Graph the offset on the chart.

8. _____ Is this level of control acceptable?

Name: _____ Date: _____

Control System Principles

_____ **1.** A(n) ___ is a control system in which compressed air is used to provide power for the control system.

    A. self-contained control system
    B. pneumatic control system
    C. electric control system
    D. automated control system

_____ **2.** The control of the environment in commercial buildings is based on ___ and ___.

    A. comfort; cost
    B. energy efficiency; comfort
    C. energy efficiency; size of building
    D. all of the above

_____ **3.** ___ is a designation of the contaminants present in the air of a building space.

T   F **4.** A self-contained control system is a control system that requires an external power supply.

_____ **5.** The fluid-filled element of a self-contained control system is known as a ___.

_____ **6.** A(n) ___ is a control system in which the power supply is low voltage (24 VAC) or line voltage (120 VAC or 220 VAC) from a step-down transformer that is wired into the building power supply.

_____ **7.** Electric control systems commonly use a(n) ___ device to control the flow of electricity in a circuit.

    A. mechanical
    B. pneumatic
    C. solid-state
    D. electronic

_____ **8.** A(n) ___ is a sensing device that consists of two different metals joined together.

_____ **9.** ___ are not often used for analog (proportional) control because only digital (ON/OFF) control devices are used.

T   F **10.** A bleedport is an orifice that allows a small volume of air to be expelled to the atmosphere.

_____ **11.** An electronic control system is a control system in which the power supply is ___ or less.

    A. 18 VDC

    B. 24 VDC

    C. 48 VDC

    D. 115 VDC

_____ **12.** The resistive bridge circuit output value is ___ when the system is at setpoint.

_____ **13.** The disadvantage of ___ is that they may require special diagnostic tools and procedures.

_____ **14.** A(n) ___ is a control system that uses digital solid-state components.

    A. electric control system

    B. electronic control system

    C. pneumatic control system

    D. automated control system (building automation system)

_____ **15.** A(n) ___ is a control system in which the duct pressure developed by the fan system is used as the power supply.

_____ **16.** ___ control devices are wired directly to the controlled devices such as compressors, fans, and pumps because both are at the same voltage level.

T   F   **17.** A system-powered control system is a control system that uses multiple control technologies.

_____ **18.** In hybrid control systems, ___ are used as an interface between different control system technologies.

    A. actuators

    B. controllers

    C. transducers

    D. power heads

_____ **19.** ___ are devices that are normally located between the transmitters or controllers and the controlled device.

_____ **20.** The most common application of ___ is the retrofit of an automated control system to a pneumatic control system.

Self-Contained Control System

_____ **1.** adjustment stem

_____ **2.** refrigerant outlet

_____ **3.** diaphragm

_____ **4.** fluid-filled element controls flow of refrigerant

_____ **5.** valve open

_____ **6.** liquid refrigerant inlet

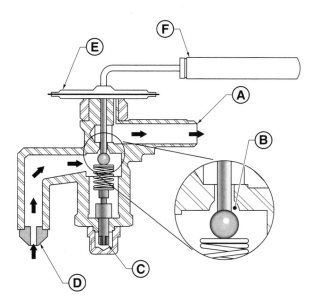

Electronic Control System

_____ **1.** electronic heating valve

_____ **2.** 18 VDC out

_____ **3.** electronic room thermostat

_____ **4.** DC power supply

_____ **5.** 24 VAC in

_____ **6.** electronic cooling valve

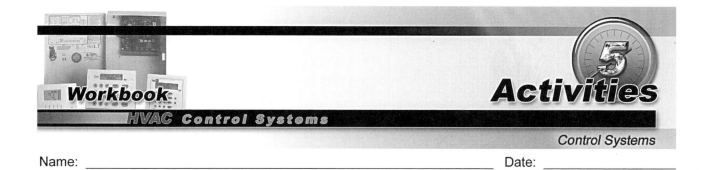

Name: _____   Date: _____

Activity 5-1. Control System Identification

A new firm has taken over the service contract on a building. Before service can be performed, knowledge of the control system must be obtained. Use the air handling unit drawing to answer the questions.

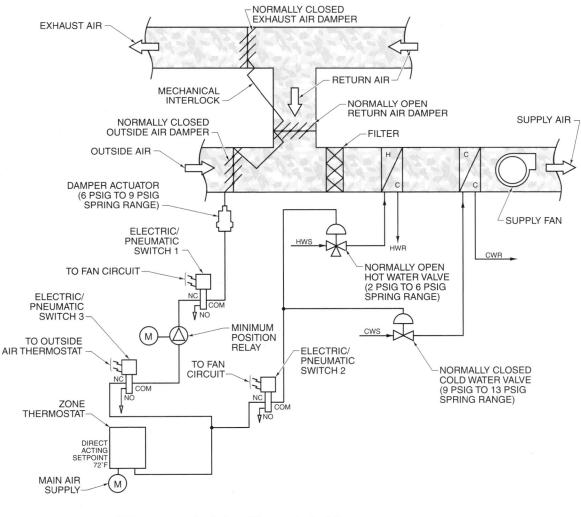

1. _____   What type of air handling unit is it?

2. _____   Based on the drawing, what type of control system is it?

Activity 5-2. Control System Device Identification

Use the pneumatic control system drawing as a guide to answer the questions on the air handling unit drawing from Activity 5-1.

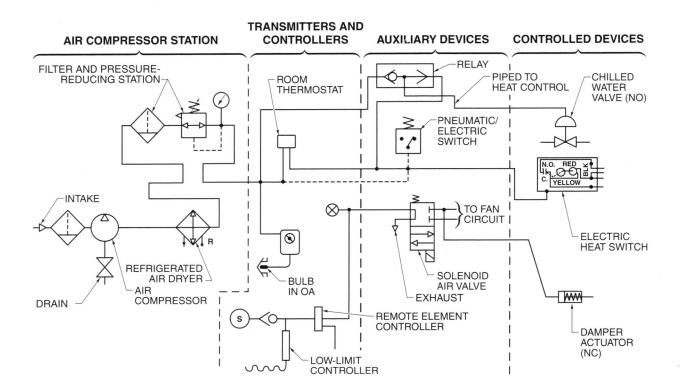

1. List the transmitter(s)/controller(s) in the air handling unit.

_____

_____

2. List the auxiliary device(s) in the air handling unit.

_____

_____

3. List the controlled device(s) in the air handling unit.

_____

_____

4. _____ The setpoint of the room thermostat is ___°F.

5. _____ What device on the unit reduces indoor air quality problems?

6. _____ Is a humidity control device included on the unit?

7. What items require replacement on a regular basis?

_____

_____

8. _____ What type of water is used in the unit?

9. What devices supply water to the unit?

_____

_____

10. _____ What device moves air in the system?

Name: _____ Date: _____

Control System Principles

T   F   **1.** An air compressor is a device that takes air from the atmosphere and compresses it to increase its pressure.

T   F   **2.** Oil carryover is compressor lubricating oil that leaks by the piston rings and is carried into the compressed air system.

_____ **3.** A(n) ___ is a unit of measure equal to .000039″.

T   F   **4.** A dessicant air drier is commonly located in a metal housing at the intake of the air compressor.

T   F   **5.** An automatic drain is a device that opens and closes automatically at a predetermined interval to drain moisture from the receiver.

_____ **6.** ___ is the adhesion of a gas or liquid to the surface of a porous material.

_____ **7.** A(n) ___ is a valve that restricts and/or blocks downstream air flow.

_____ **8.** A(n) ___ valve is a device at the compressor station that prevents excessive pressure from building up by venting air to the atmosphere.

_____ **9.** A(n) ___ test is an air compressor performance test that measures the percentage of time that a compressor runs to maintain a supply of compressed air to the control system.

_____ **10.** An ___ filter is a device at the discharge side of the compressor that removes oil droplets from air by forcing compressed air to change direction quickly.

      A. oil removal
      B. air line
      C. intake air
      D. oil saturated

_____ **11.** A(n) ___ air compressor consists of two air compressors and two electric motors on one common receiver.

T   F   **12.** A compressor alternator is a pressure switch that determines which compressor is the primary (lead) compressor and which compressor is the backup (lag) compressor.

T   F   **13.** A lead/lag switch is a device that operates one compressor during one pumping cycle and the other compressor during the next pumping cycle.

T   F   **14.** A screw compressor uses a pair of rotating interlocking rotors to compress air.

_____ **15.** A(n) ___ compressor is a positive-displacement compressor that has multiple vanes located in an offset rotor.

T   F   **16.** A dynamic compressor uses centrifugal force to move air.

_____ **17.** A(n) ___ drier is a device that uses refrigeration to lower the temperature of compressed air.

    A. refrigerated air
    B. desiccant
    C. charcoal
    D. air line

_____ **18.** A(n) ___ is a device that is opened and closed manually to drain moisture from the receiver.

_____ **19.** In general, air compressors should use ___ oil.

    A. synthetic
    B. detergent
    C. non-detergent
    D. water-based

T   F   **20.** If compressed air that contains moisture is exposed to low outside temperatures, air lines may freeze and split.

Oil Removal Filters

_____ **1.** automatic drain

_____ **2.** bowl guard

_____ **3.** filtration element

_____ **4.** body

_____ **5.** bowl O-ring

_____ **6.** drain O-ring

_____ **7.** nut

_____ **8.** transparent bowl

_____ **9.** differential pressure indicator

_____ **10.** gasket

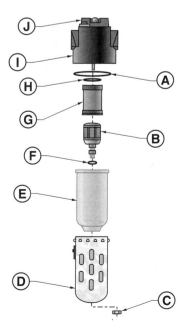

Air Compressor Preventive Maintenance Procedure

Identify the procedures shown.

_____ 1. Properly replace belt guard and turn power ON to compressor.

_____ 2. Drain tank. Check volume of oil in water.

_____ 3. Turn power OFF to compressor. Follow lockout/tagout procedures.

_____ 4. Check oil level in crankcase.

_____ 5. Remove belt guard, check belt for cracking, glazing, and tension.

_____ 6. Manually operate safety relief valves.

(A)

(B)

(C)

(D)

(E)

(F)

Air Compressor Schematic Symbol Identification

_____ **1.** filter/separator with manual drain

_____ **2.** automatic drain

_____ **3.** pressure switch

_____ **4.** intake air filter

_____ **5.** safety relief valve

_____ **6.** pressure regulator with gauge

_____ **7.** filter/separator with automatic drain

_____ **8.** air compressor

_____ **9.** air drier

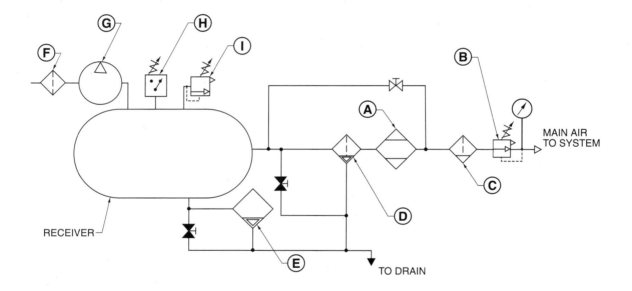

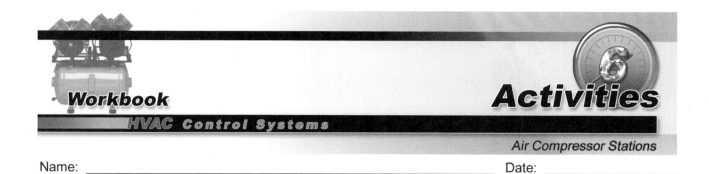

Name: _____ Date: _____

Activity 6-1. Air Compressor Performance

A control air compressor under normal load is timed as running from 7:10 AM to 7:15 AM. The compressor then starts again at 7:25 AM.

1. _____ The compressor run time is ___%.

2. _____ Is the compressor run time within normal parameters?

3. _____ The compressor has ___ starts per hour.

4. _____ Is the compressor number of starts per hour within normal parameters?

Activity 6-2. Log Sheet Completion

Complete the air compressor preventive maintenance log.

1.

AIR COMPRESSOR PREVENTIVE MAINTENANCE LOG										
Air Compressor Location:	Number:	Date:	On-Time:	Off-Time:	Run-Time:	Starts/Hour:	PM Performed?	Problems:	Notes:	Initials:
Admin Bldg	1	11/7	5 min	10 min			Yes	None		
Admin Bldg	2	11/7	4 min	4 min			Yes	None		
Gym	1	11/7	4 min	10 min			Yes	None		

Activity 6-3. Control Air System Component Connection

Connect the devices in the control air system.

1.

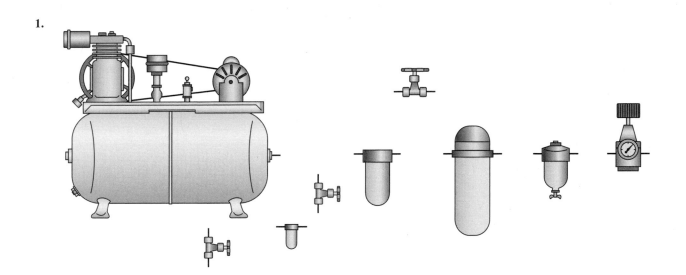

Name: _____ Date: _____

Control System Principles

_____ 1. A(n) ___ is a device that accepts a signal from a controller and causes a proportional mechanical motion to occur.

_____ 2. A(n) ___ is a flexible device that transmits the force of the incoming air pressure to the piston cup and then to the spring and shaft assembly.

    A. diaphragm
    B. piston cup
    C. aluminum end cap
    D. spring and shaft assembly

_____ 3. A(n) ___ is a device that transfers the force generated by the air pressure against the diaphragm to the spring and shaft assembly.

T   F   4. Spring range is the difference in pressure at which an actuator shaft moves and stops.

_____ 5. Spring range ___ is a condition in which actuators with different spring ranges interfere with each other.

_____ 6. A(n) ___ is an adjustable metal blade or set of blades used to control the flow of air.

_____ 7. ___ blade dampers are the most common damper used in HVAC systems.

    A. Round
    B. Opposed
    C. Parallel
    D. Elliptical

_____ 8. A(n) ___ is a device that controls the flow of fluids in an HVAC system.

_____ 9. A(n) ___ valve has two pipe connections.

_____ 10. A(n) ___ valve does not allow fluid to flow when the valve is in its normal position.

_____ 11. A(n) ___ is a valve component that consists of a metal shaft, normally made of stainless steel, that transmits the force of the actuator to the valve plug.

T   F   12. A normally closed valve allows fluid to flow when the valve is in its normal position.

_____ **13.** A(n) ___ valve is a three-way valve that has two inlets and one outlet.

    A. equal percentage
    B. diverting
    C. butterfly
    D. mixing

_____ **14.** A(n) ___ valve is a valve in which the flow through the valve is equal to the amount of valve stroke.

    A. butterfly
    B. linear
    C. quick-opening
    D. diverting

T   F   **15.** Steam valves are always two-way valves because steam does not require a return to the steam header or boiler.

_____ **16.** ___ characteristics is the relationship between the valve stroke and flow through the valve.

_____ **17.** Valve ___ is the relationship between the maximum flow and the minimum controllable flow through a valve.

    A. turn down ratio
    B. shut-off rating
    C. control ratio
    D. shut-down rating

_____ **18.** ___ is a bulk deformable material or one or more mating deformable elements reshaped by manually adjustable compression.

_____ **19.** Valve ___ is performed when a valve is mechanically damaged or the internal parts are worn by foreign material or wire drawing.

T   F   **20.** Most air handling systems use normally open outside air dampers because the outside air causes excessively cold conditions in the winter and excessively hot and humid conditions in the summer.

Dampers

_____ **1.** bottom seal

_____ **2.** end seal

_____ **3.** damper blade

_____ **4.** damper blade seal

_____ **5.** damper blade pin

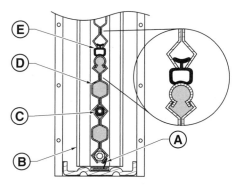

Valve Rebuild Kit

_____ **1.** stem and disc holder assembly

_____ **2.** follower spring

_____ **3.** packing nut

_____ **4.** packing lubricant

_____ **5.** follower

_____ **6.** valve body

_____ **7.** plug

_____ **8.** disc spring

_____ **9.** packing gland

_____ **10.** packing

_____ **11.** bonnet

_____ **12.** disc

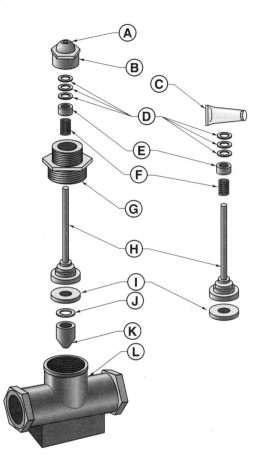

Damper Configurations

_____ **1.** opposed blade

_____ **2.** round blade

_____ **3.** parallel blade

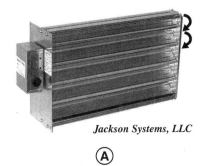

Jackson Systems, LLC

Ⓐ

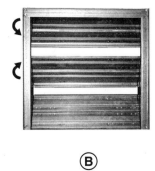

Ⓑ

Jackson Systems, LLC

Ⓒ

HVAC Control Valve Components

_____ **1.** valve body

_____ **2.** packing spring

_____ **3.** packing gland

_____ **4.** packing

_____ **5.** air fitting

_____ **6.** piston cup

_____ **7.** stem

_____ **8.** valve plug

_____ **9.** seat

_____ **10.** disc

_____ **11.** flange

_____ **12.** stem and plug assembly

_____ **13.** packing nut

_____ **14.** disc holder

_____ **15.** valve port

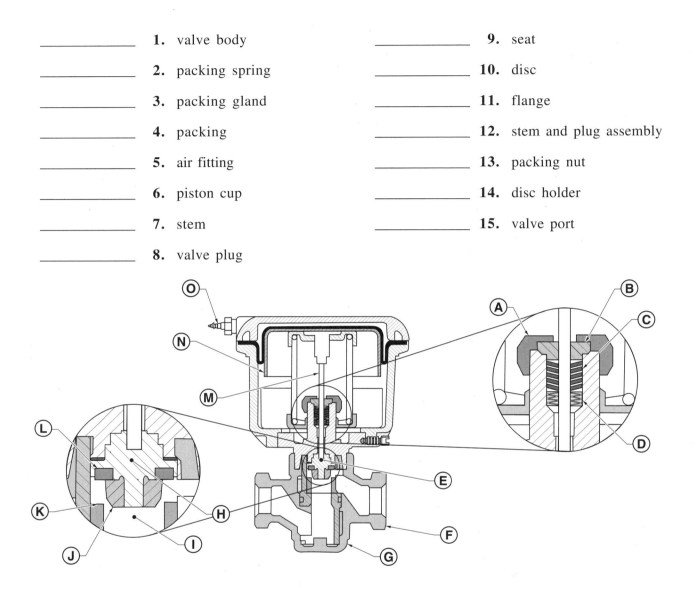

Valves

_____ **1.** packing gland

_____ **2.** packing nut

_____ **3.** bonnet

_____ **4.** packing

_____ **5.** valve stem

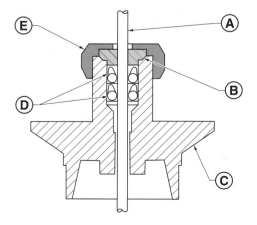

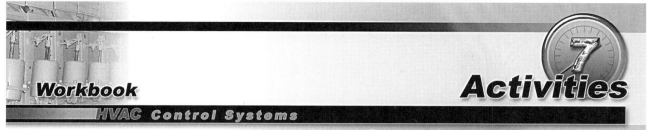

Name: _____   Date: _____

Activity 7-1. Valve Data Sheet

Answer the questions using the valve data sheet.

VALVE DATA SHEET

Product		VP8200 Series Bronze Control Valves
Service*		Hot Water, Chilled Water, Glycol Solutions, or Steam for HVAC Systems
Valve Body Size/Cv	1/2 in.	0.73, 1.8, and 4.6
	3/4 in.	7.3
	1 in.	11.6
	1-1/4 in.	18.5
	1-1/2 in.	28.9
	2 in.	46.2
Valve Stroke		5/16 in. for 1/2 and 3/4 in. Valves
		1/2 in. for 1 and 1-1/4 in. Valves
		3/4 in. for 1-1/2 and 2 in. Valves
Valve Body Rating	Steam Brass Trim	38 psig Saturated Steam at 284°F
Valve Assembly	SS Trim	100 psig Saturated Steam at 284°F
Maximum Allowable	Water Brass Trim	400 psig Up to 150°F, Decreasing to 365 psig at 248°F
Pressure/Temperature	SS Trim	400 psig Up to 150°F, Decreasing to 308 psig at 338°F
Inherent Flow Characteristics		Equal Percentage: N.O./PDTC and N.C./PDTO Valves
		Linear: Three-Way Mixing Valves
Rangeability**		25:1 for All Sizes
Spring Range		3 to 6 psig: 3 to 7 psig for MP6000 Series
Pneumatic Actuators		4 to 8 psig
		9 to 13 psig
Maximum Recommended	Steam Brass Trim	15 psig for All Valve Sizes
Operating Pressure Drop	SS Trim	100 psig for All Valve Sizes
	Water All Trim	35 psig for 1/2 through 1-1/2 in. Valves
		30 psig for 1-1/2 and 2 in. Valves
Maximum Actuator Supply Pressure		25 psig Maximum

*Proper water treatment is recommended.
** Rangeability is the ratio of maximum flow to minimum controllable flow.

1. _____   The ___ valve should be selected if an application requires a Cv of 25.

2. _____   The valve stroke is ___″.

3. _____   The three-way mixing valves have ___ flow characteristics.

4. _____   The maximum recommended operating pressure drop for a 1½″ water valve is ___ psig.

5. List all possible spring ranges for pneumatic actuators.

_____

_____

Activity 7-2. Hot Water Heat Exchanger

Answer the questions using the heat exchanger drawing.

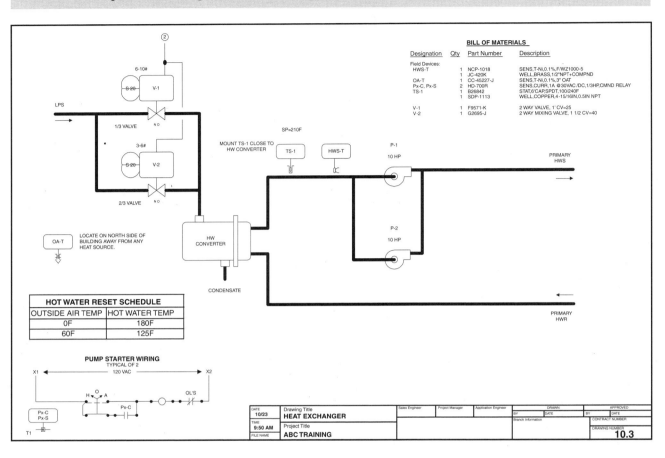

1. _____ The type of application is ___.

2. Are valves V-1 and V-2 normally open or normally closed? Why?

_____

_____

3. _____ Are valves V-1 and V-2 the same size?

4. _____ Valve V-1 is a(n) ___ valve.

5. _____ The Cv of valve V-1 is ___.

6. Why do the actuators for valve V-1 and valve V-2 have different spring ranges?

_____

_____

7. _____ According to the hot water reset schedule, if the outside air temperature is 60°F, the hot water setpoint is ___°F.

8. Would valve V-2 tend to be open or closed at a 55°F outside air temperature? Why?

_____

_____

9. If the pumps were 20 HP instead of 10 HP and delivered twice the amount of water and flow rate, would it affect the valve sizing? Why or why not?

_____

_____

Activity 7-3. Air Handling Unit

Answer the questions using the air handling unit drawing.

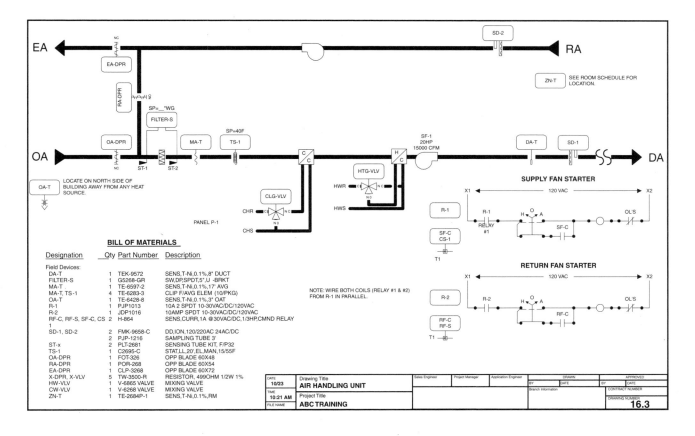

1. _____ The system contains ___ (number) dampers.

2. List each damper designation and size.

_____

_____

3. Is the outside air (OA) damper normally open or normally closed? Why?

---

4. _____ The return air damper is a(n) ___ type of damper.

5. _____ The designation of the heating valve is ___.

6. _____ The description of the HTG-VLV is ___.

7. _____ The designation of the cooling valve is ___.

8. _____ The description of the CLG-VLV is ___.

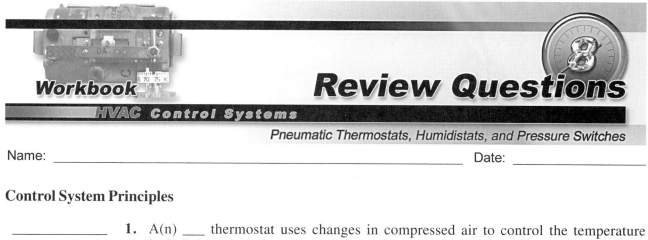

Name: _____ Date: _____

Control System Principles

_____ 1. A(n) ___ thermostat uses changes in compressed air to control the temperature in individual rooms inside a commercial building.

_____ 2. A ___ is a sensing device that consists of two different metals joined together.
   A. setpoint
   B. bimetallic element
   C. baseplate
   D. bleedport

_____ 3. ___ pressure is the pressure in the air line that is piped from the thermostat to the controlled device.

_____ 4. A ___ thermostat changes the air pressure to a valve or damper actuator by changing the amount of air that is expelled to the atmosphere.
   A. bleed-type
   B. pilot bleed
   C. single-temperature-setpoint
   D. limit

_____ 5. A(n) ___ application is an application in which the thermostat is located in a return air slot that is integral to a light fixture.

_____ 6. A(n) ___ is a fixed orifice that meters air flow through a port and allows fine output pressure adjustments and precise circuit control.

_____ 7. A(n) ___ thermostat increases the branch line pressure as the building space temperature increases and decreases the branch line pressure as the building space temperature decreases.

T   F   8. The metallic element in a pilot bleed thermostat acts like an amplifier.

_____ 9. A(n) ___ is used to override a thermostat to the day temperature during the night mode.

T   F   10. Winter/summer thermostats have two bimetallic elements.

_____ 11. A(n) ___ is the range between two temperatures in which no heating or cooling takes place.

_____ 12. A(n) ___ is an orifice that allows a small volume of air to be expelled to the atmosphere.

_____ **13.** A ___ thermostat is a pneumatic thermostat that has a third air connection piped to either a manual air regulator or an outside air transmitter.

    A. deadband pneumatic
    B. master/submaster
    C. limit
    D. chilled-water

_____ **14.** A(n) ___ is a controller that uses compressed air to open or close a device which maintains a certain humidity level inside a duct or area.

_____ **15.** A(n) ___ is a controller that maintains a constant air pressure in a duct or area.

_____ **16.** A(n) ___ relay causes the operation of a thermostat to change between two or more modes such as day/night.

_____ **17.** A winter/summer thermostat is a ___ thermostat that changes the setpoint and action of the thermostat from the winter (heating) to the summer (cooling) mode.

    A. limit
    B. deadband pneumatic
    C. pilot bleed
    D. single-temperature-setpoint

T   F   **18.** A dual-duct air handling unit supplies both hot and cold air to a sheet metal mixing box.

_____ **19.** The ___ element in a pneumatic humidistat accepts moisture from, or rejects moisture to, the surrounding air.

    A. hygroscopic
    B. hydropic
    C. butyrate
    D. diaphragm

_____ **20.** A(n) ___ thermostat is a pneumatic thermostat that maintains a temperature above or below an adjustable setpoint.

Winter/Summer Thermostats

_____ **1.** control line test gauge

_____ **2.** direct-acting strip

_____ **3.** sensitivity slider

_____ **4.** reverse-acting strip

_____ **5.** heating adjusting screw

_____ **6.** cooling adjusting screw

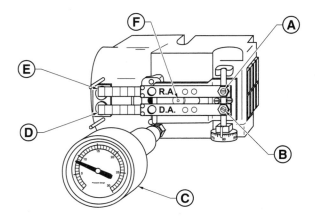

Day/Night Thermostat Calibration

Identify the procedures shown.

_____ **1.** Adjust night setpoint dial to desired night temperature.

_____ **2.** Adjust day output to midpoint of actuator spring range.

_____ **3.** Adjust both setpoint dials to ambient temperature.

_____ **4.** Change main air pressure to night pressure.

_____ **5.** Adjust day setpoint dial to desired day temperature.

_____ **6.** Change main air pressure to day pressure.

_____ **7.** Remove cover and install pressure gauge.

_____ **8.** Adjust night output to midpoint of actuator spring range.

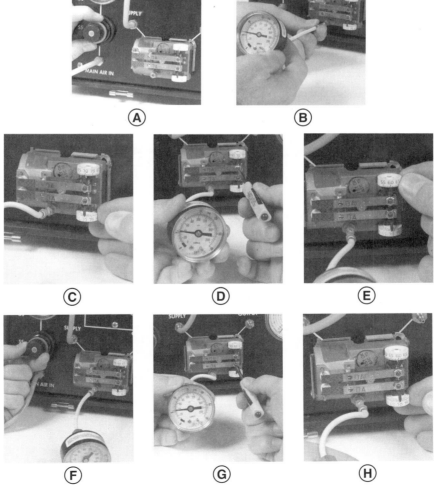

Limit Thermostats

_____ **1.** electrical/pneumatic switch

_____ **2.** limit thermostat

_____ **3.** pressure switch

_____ **4.** fan

_____ **5.** actuator

_____ **6.** room thermostat

_____ **7.** N/O steam valve

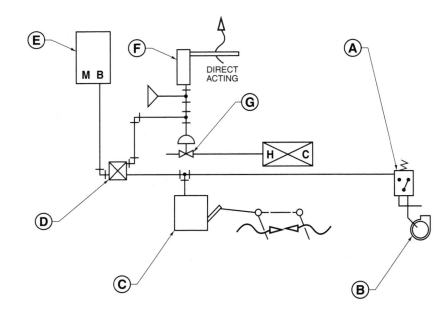

Pilot Bleed Single-Temperature Thermostat Calibration

Identify the procedures shown.

_____ **1.** Adjust branch line pressure to midpoint of actuator spring range.

_____ **2.** Remove pressure gauge and thermostat. Reinstall cover.

_____ **3.** Remove thermostat cover.

_____ **4.** Install pressure gauge and thermometer.

_____ **5.** Check repeatability by turning setpoint dial ±2°F.

_____ **6.** Turn setpoint dial to ambient temperature.

_____ **7.** Turn setpoint dial to desired setpoint.

_____ **8.** Wait 5 min. Measure air temperature near thermostat.

(A)

(B)

(C)

(D)

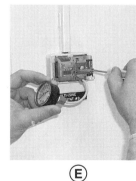
(E)

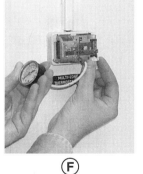
(F)

(G)

(H)

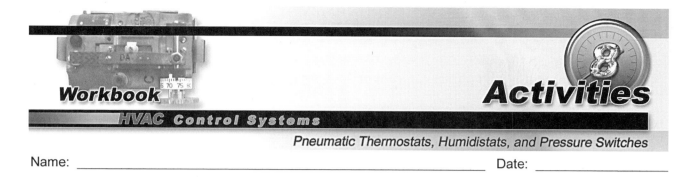

Name: _____ Date: _____

Activity 8-1. Troubleshooting VAV Cooling with Finned-Tube Hot Water Heating System

The occupant in a room has called indicating that the room temperature is too hot. It is the first warm spell of the year and the cooling equipment has just started running. The temperature in the room is 80°F. The setpoint of the thermostat is 74°F. Use the drawing to answer the questions.

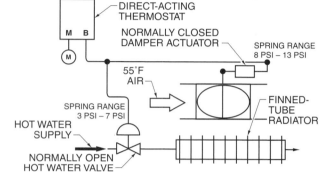

1. _____ Is the thermostat in the diagram direct- or reverse-acting?

2. _____ As the room temperature increases, should the output pressure increase or decrease?

The cover of the thermostat is removed and an output pressure gauge installed. The output pressure gauge reads 20 psig.

3. What are three possible problems?

_____

_____

The output pressure can be adjusted from 0 psig to 20 psig easily. As the output pressure is adjusted lower, the supply of 55°F air from the duct seems to decrease and water noise increase, indicating the fin tube coil is heating up.

4. Is the damper or valve stuck? Why or why not?

_____

5. If the mechanical system is working, should the calibration of the thermostat be checked? If so, give the procedure.

_____

_____

Activity 8-2. Troubleshooting Finned-Tube Steam Heating System

The occupant in a room has called indicating that the room temperature was fine until an hour ago but is now too hot. The temperature in the room is 80°F. The setpoint of the thermostat is 72°F. Use the drawing to answer the questions.

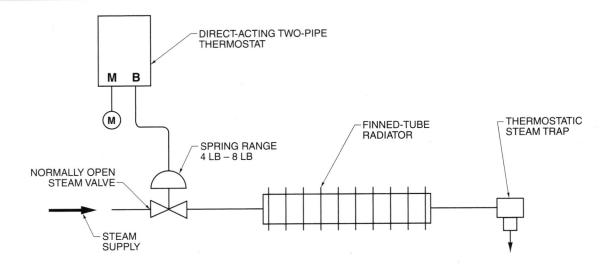

1. _____ Is the thermostat in the diagram direct- or reverse-acting?

2. _____ As the room temperature increases, should the output pressure increase or decrease?

3. _____ If the thermostat action was the opposite, should the output pressure increase or decrease on an increase in room temperature?

The cover of the thermostat is removed and an output pressure gauge installed. The output pressure gauge reads 0 psig.

4. What are three possible problems?

_____

_____

_____

Regardless of the amount the output pressure is adjusted, it remains at 0 psig.

5. _____ What tool is needed to check the operation of the actuator?

The access door to the finned-tube radiator is opened and steam at full flow is observed through the valve. A squeeze bulb is connected to the actuator by removing the air line from the thermostat. An attempt is made to pump air into the actuator and close it, but regardless of the amount of pumping, the actuator pressure does not increase.

6. What is the problem?

_____

_____

7. Can it be fixed? If so, give the procedure.

_____

_____

8. How can the actuator be tested to ensure proper operation and no callbacks?

_____

_____

Name: _____   Date: _____

Control System Principles

_____ **1.** A(n) ___ is a device that senses temperature, pressure, or humidity and sends a proportional (3 psig to 15 psig) signal to a controller.

_____ **2.** A(n) ___ device uses a small amount of the compressed air supply (restricted main air).

_____ **3.** A(n) ___ device uses the full volume of compressed air available.

T   F   **4.** All pneumatic transmitters have a range and span.

_____ **5.** Transmitter ___ is the output pressure change that occurs per unit of measured variable change.

      A. sensitivity
      B. range
      C. span
      D. gauge pressure

_____ **6.** A room ___ is a transmitter used in applications that require a receiver controller to measure and control the temperature in an area.

_____ **7.** A(n) ___ temperature transmitter is a pneumatic transmitter that uses a long tube filled with a liquid or gas to sense duct temperature.

      A. rod-and-tube
      B. inert-element
      C. averaging element
      D. bulb-type

_____ **8.** A ___ transmitter is a device that measures the amount of moisture in the air compared to the amount of moisture the air could hold if it were saturated.

      A. room humidity
      B. bulb-type
      C. duct humidity
      D. pneumatic humidity

T   F   **9.** A duct humidity transmitter is a device used to sense the pressure due to air flow in a duct or water flow through a pipe.

_____ **10.** A(n) ___ is a device that senses static pressure and total pressure in a duct.

_____ **11.** A(n) ___ transmitter is a device mounted in a duct that senses the static pressure due to air movement.

    A. duct pressure
    B. pipe pressure
    C. humidity
    D. inert-element

_____ **12.** ___ transmitters are used in HVAC piping systems to sense the pressure in water distribution systems.

_____ **13.** ___ transmitters are used to sense the relative humidity in a building space and send a 3 psig to 15 psig signal to a receiver controller.

_____ **14.** A(n) ___ transmitter is a pneumatic transmitter that is used to sense outside air temperatures in through-the-wall applications.

    T   F   **15.** All pneumatic transmitters have a linear output.

_____ **16.** A(n) ___ transmitter is a pneumatic transmitter that uses a high-quality metal rod with precise expansion and contraction characteristics as the sensing element.

_____ **17.** A(n) ___ is a spring or thin piece of metal that measures the movement of a sensing element.

_____ **18.** ___ transmitters have sensing elements 6″ to 12″ long and are used to sense air temperature in air handling systems.

_____ **19.** ___ transmitters have sensing elements approximately 4″ long and are used to sense water temperatures in hot water, chilled water, or cooling tower systems.

_____ **20.** A duct ___ allows the air from the air stream to be forced across the humidity-sensing element.

    A. room humidity transmitter
    B. manometer
    C. sampling tube kit
    D. bulb transmitter

Rod-and-Tube Temperature Transmitters

_____ **1.** brass tube

_____ **2.** bleed port

_____ **3.** metal rod

_____ **4.** sensor line port

_____ **5.** spring

_____ **6.** flapper/nozzle assembly

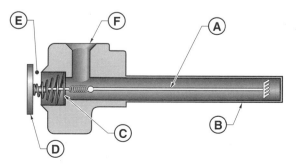

Duct Pressure Transmitters

_____ **1.** pitot tube

_____ **2.** discharge air

_____ **3.** normally open damper

_____ **4.** receiver controller

_____ **5.** duct pressure transmitter

_____ **6.** duct

_____ **7.** static pressure

_____ **8.** pneumatic damper actuator

_____ **9.** static pressure probe

_____ **10.** reference pressure probe

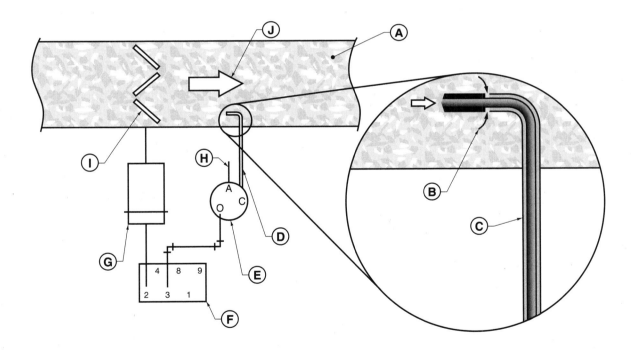

Inert-Element Temperature Transmitters

_____ **1.** sensing element shield

_____ **2.** inert section

_____ **3.** inert-element temperature transmitter

_____ **4.** long-rod section (sensing element)

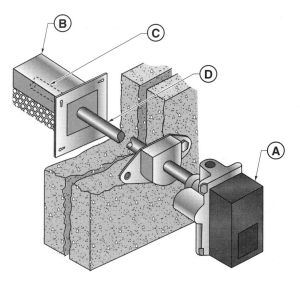

Pneumatic Transmitters

_____ **1.** coil

_____ **2.** pump

_____ **3.** sun shield

_____ **4.** reset controller

_____ **5.** boiler

_____ **6.** hot water sensor

_____ **7.** three-way mixing valve

_____ **8.** outside air pneumatic transmitter

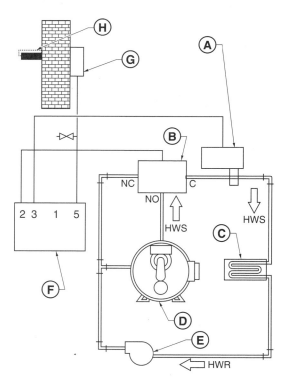

Name: _____ Date: _____

MULTIZONE AIR HANDLING UNIT

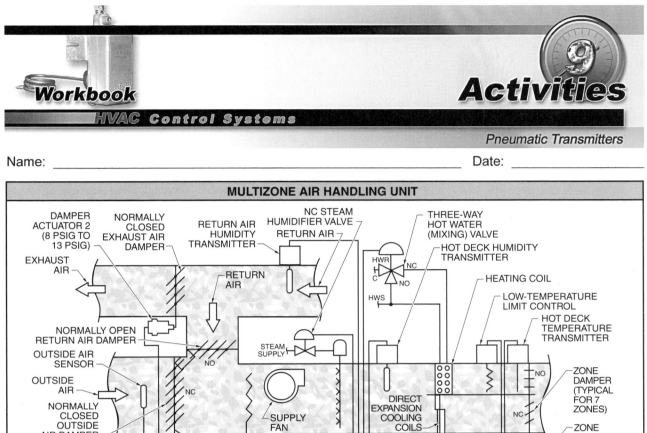

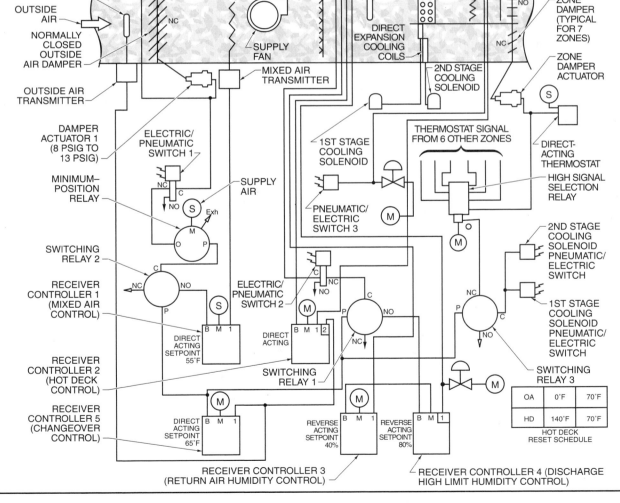

OA	0°F	70°F
HD	140°F	70°F

HOT DECK
RESET SCHEDULE

AVERAGING ELEMENT TRANSMITTERS

Order No.	Sensing Range*	Element Length[†]	Mounting
MA825B 1003	–40 to 160	.15	Duct
MA825B 1011	–40 to 160	27[‡]	Wall
MA825B 1029	40 to 240	15	Well
MA825B 1037	–40 to 160	15	Well
MA825B 1045	–40 to 160	7	Duct
MA825B 1243	–20 to 80	15	Duct
MA825B 1250	–20 to 80	27[‡]	Wall
MA825B 1268	40 to 240	15	Duct
Order No.	Sensing Range*	Element Length[§]	Mounting
MA826B 1044	0 to 200	18½	Duct
MA826B 1051	0 to 200	8⅞	Duct
MA826B 1077	25 to 125	18½	Duct
MA826B 1085	25 to 125	8⅞	Duct

* Non-adjustable in °F
† in in.
‡ Active element 15″, inert section 12″
§ in ft

TRANSMITTER CHARACTERISTICS

Temperature Range*	Span*	Sensitivity[†]
40 – 65 60 – 85	25	.48
50 – 100	50	.24
0 – 100 20 – 120 50 – 150	100	.12
40 – 240 –40 – 160 200 – 400	200	.06

* in °F
† in psi/°F

TEMPERATURE AND RELATIVE HUMIDITY vs. TRANSMITTER OUTPUT PRESSURE

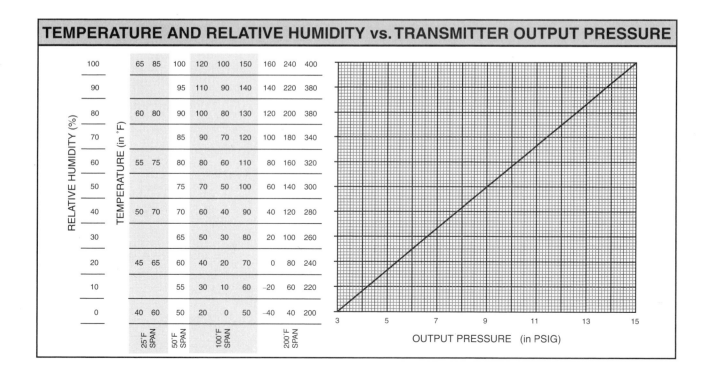

Activity 9-1. Mixed Air Pneumatic Transmitter Selection

Locate the mixed air transmitter on the multizone air handling unit drawing and label it T-1 for your reference. The mixed air transmitter is installed inside a large air duct and must obtain an accurate air temperature sample. The expected value of the mixed air temperature is approximately 55°F.

1. Using the Averaging Element Transmitters chart, what possible transmitter models may be selected?

   _____

   _____

2. What factors determine the selection?

   _____

   _____

3. _____ The most appropriate transmitter temperature range is ___°F.

Use the Transmitter Characteristics chart to determine the transmitter span and sensitivity.

4. _____ The selected transmitter span is ___°F.

5. _____ The selected transmitter sensitivity is ___ psig/°F.

Use the Temperature and Relative Humidity vs. Transmitter Output Pressure chart to determine the output pressure.

6. _____ The output pressure of the selected transmitter at 60°F and 40% rh is ___ psig.

7. _____ The output pressure of the selected transmitter at 40°F and 20% rh is ___ psig.

8. _____ The output pressure of the selected transmitter at 80°F and 30% rh is ___ psig.

Activity 9-2. Hot Deck Pneumatic Transmitter Selection

Locate the hot deck transmitter on the multizone air handling unit drawing and label it T-2 for your reference. The hot deck transmitter is installed inside a small air duct.

1. _____ Using the hot deck reset schedule, what is the expected hot deck temperature range?

2. Using the Averaging Element Transmitters chart, what possible transmitter models may be selected?

   _____

   _____

3. What factors determine the selection?

_____

_____

4. _____ The most appropriate transmitter temperature range is ___°F.

Use the Transmitter Characteristics chart to determine the transmitter span and sensitivity.

5. _____ The selected transmitter span is ___°F.

6. _____ The selected transmitter sensitivity is ___ psig/°F.

Use the Temperature and Relative Humidity vs. Transmitter Output Pressure chart to determine the output pressure.

7. _____ The output pressure of the selected transmitter at 60°F and 50% rh is ___ psig.

8. _____ The output pressure of the selected transmitter at 100°F and 30% rh is ___ psig.

9. _____ The output pressure of the selected transmitter at 150°F and 55% rh is ___ psig.

Activity 9-3. Outside Air Pneumatic Transmitter Selection

Locate the outside air temperature transmitter on the multizone air handling unit drawing and label it T-3 for your reference. The outside air temperature transmitter is installed through the wall into the outside air.

1. Using the Averaging Element Transmitters chart, what possible transmitter models may be selected?

_____

_____

2. What factors determine the selection?

_____

_____

3. _____ Based on the local climate, what is the expected outside air temperature range?

4. _____ The most appropriate transmitter temperature range is ___°F.

Use the Transmitter Characteristics chart to determine the transmitter span and sensitivity.

5. _____ The selected transmitter span is ___°F.

6. _____ The selected transmitter sensitivity is ___ psig/°F.

Use the Temperature and Relative Humidity vs. Transmitter Output Pressure chart to determine the output pressure.

7. _____ The output pressure of the selected transmitter at 35°F and 35% rh is ___ psig.

8. _____ The output pressure of the selected transmitter at 75°F and 55% rh is ___ psig.

9. _____ The output pressure of the selected transmitter at 95°F and 45% rh is ___ psig.

Name: _____ Date: _____

Control System Principles

_____ 1. A(n) ___ is a device which accepts one or more input signals from pneumatic transmitters and produces an output signal based on its calibration.

_____ 2. A(n) ___ is a device that consists of a controller mounted to a duct or pipe and that is connected by a capillary tube to a bulb that is inserted into the duct or pipe.

    A. single-input receiver controller
    B. dual-input receiver controller
    C. remote bulb controller
    D. all of the above

_____ 3. A(n) ___ receiver controller is a controller in which the controller output is determined by the relationship of mechanical pressures.

T   F 4. A force-balance controller uses vector analysis to arrive at an output.

_____ 5. A pneumatic ___ senses temperature, pressure, or humidity and sends a proportional 3 psig to 15 psig signal to a receiver controller.

_____ 6. Symbols used to indicate that the internal restricted main air supply to a transmitter is shut OFF include ___.

    A. box around transmitter port
    B. solid circle
    C. bow tie with line in the center
    D. all of the above

_____ 7. A(n) ___ may be required to prevent undue pressure drop in extremely long piping runs (over 250′).

_____ 8. A(n) ___ controller is designed to be connected to only one transmitter and to maintain only one temperature, pressure, or humidity setpoint.

_____ 9. ___ is the number of units of controlled variable that causes an actuator to move through its entire spring range.

    A. Tubing jumper band
    B. Point adjustment
    C. Proportional band
    D. Gain

_____ **10.** ___ is the mathematical relationship between the controller output pressure change and the transmitter pressure change that causes it.

T   F **11.** Calibration of a single-input receiver controller without a simulator kit is as precise as when using a simulator.

_____ **12.** Dual-duct and multizone air handling units have a ___ configuration.

_____ **13.** Single-input receiver controllers are often used to control vital ___ systems in a building.

_____ **14.** A(n) ___ receiver controller is a receiver controller in which the change of one variable, commonly outside air temperature, causes the setpoint of the controller to automatically change (reset) to match the changing condition.

        A. single-input
        B. dual-input
        C. remote bulb
        D. externally restricted

_____ **15.** A(n) ___ is a chart that describes the setpoint changes in a pneumatic control system.

_____ **16.** ___ is a schedule in which the primary variable increases as the secondary variable increases, and decreases as the secondary variable decreases.

        A. Reset schedule
        B. Integration schedule
        C. Reverse readjustment (winter reset)
        D. Direct readjustment (summer reset)

T   F **17.** All receiver controllers should be recalibrated a minimum of once a year.

_____ **18.** ___ is the ability to adjust the controller setpoint from a remote location.

_____ **19.** ___ is a function that calculates the amount of difference between the setpoint and control point (offset) over time.

_____ **20.** Controllers that add integration are referred to as ___ controllers.

Hot Water Control

_____ **1.** steam trap

_____ **2.** steam supply

_____ **3.** steam valve

_____ **4.** heat exchanger

_____ **5.** direct-acting controller

_____ **6.** pump

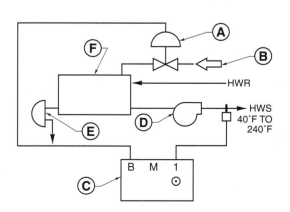

Receiver Controllers

_____	**1.** master test point
_____	**2.** remote or local setpoint test point
_____	**3.** gain adjustment dial
_____	**4.** controlled variable
_____	**5.** readjustment side
_____	**6.** supply
_____	**7.** output
_____	**8.** controlled variable test point

_____	**9.** internal supply
_____	**10.** setpoint dial
_____	**11.** ratio adjustment dial
_____	**12.** master
_____	**13.** ratio selection jumper
_____	**14.** setpoint (remote)
_____	**15.** internal supply test point for factory use only

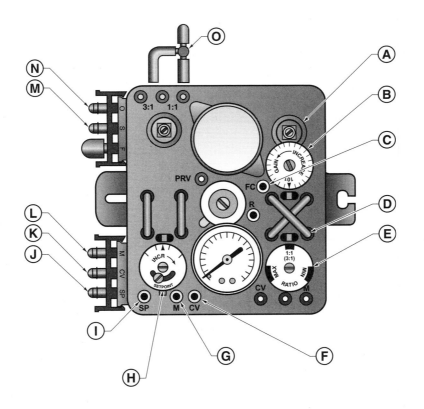

Single-Input Receiver Controller Calibration

Identify the procedures shown.

_____ **1.** Adjust output pressure of controller to midpoint of actuator spring range.

_____ **2.** Disconnect transmitter air line and connect simulator to transmitter port.

_____ **3.** Check throttling range and gain adjustment. Readjust if necessary.

_____ **4.** Determine transmitter range, setpoint, throttling range, gain, and actuator spring range.

_____ **5.** Enter desired throttling range and gain on controller.

_____ **6.** Remove simulator, reattach air lines, and check actual operation.

_____ **7.** Adjust simulator pressure-regulating valve to desired setpoint.

(A)

(B)

(C)

(D)

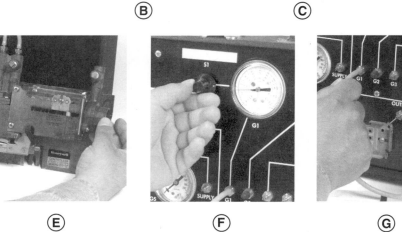

(E)

(F)

(G)

Name: _____ Date: _____

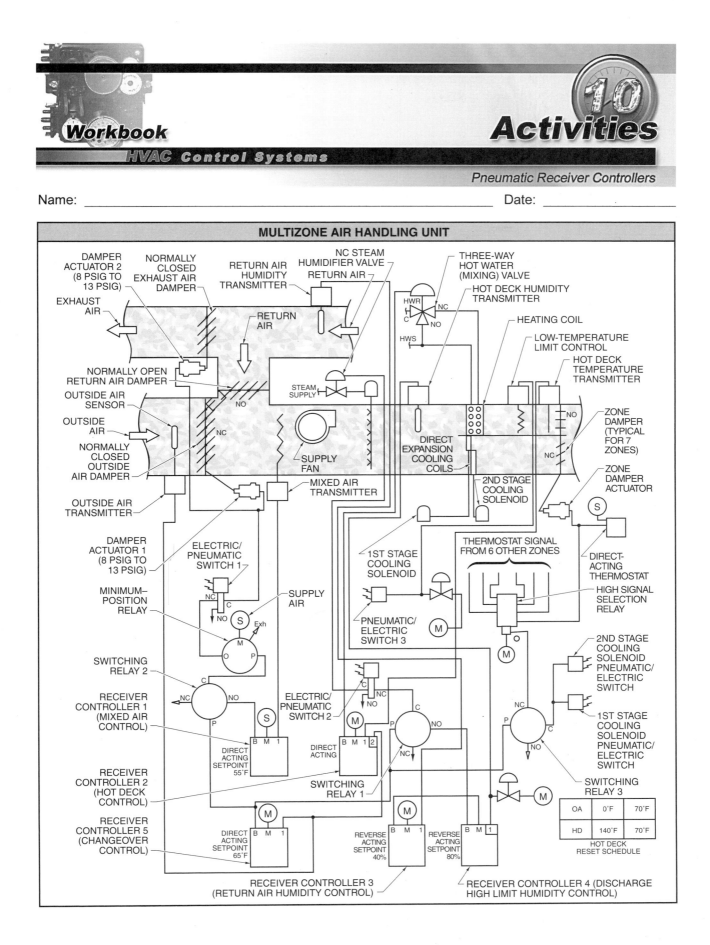

MULTIZONE AIR HANDLING UNIT

Activity 10-1. Changeover Control

Locate receiver controller 5 (changeover control) on the multizone air handling unit drawing and answer the questions.

1. _____ Is receiver controller 5 a single- or dual-input receiver controller?

2. _____ The controller setpoint is ___.

3. _____ The controller action is ___.

4. List all the devices piped to the branch line of the controller.

_____

_____

5. Is the controller internally or externally restricted? How can this be checked?

_____

_____

6. _____ At an outside air temperature of 77°F, the output pressure from the controller is ___ psig.

7. _____ At an outside air temperature of 35°F, the pressure at port P of switching relay 1 is ___ psig.

Activity 10-2. Hot Deck Control

Locate receiver controller 2 (hot deck control) on the multizone air handling unit drawing and answer the questions.

1. _____ Is receiver controller 2 a single- or dual-input receiver controller?

2. _____ The controller setpoint/reset schedule is ___.

3. _____ At a 35°F outside air temperature, the hot deck setpoint is ___°F.

4. _____ The controller action is ___.

5. _____ The controller readjustment is ___.

6. List all the devices piped to the branch line of the controller.

_____

_____

7. _____ The ___ transmitter piped to the controller is internally restricted.

Activity 10-3. Humidity Control

Locate receiver controller 3 (return air humidity control) and receiver controller 4 (discharge high limit humidity control) on the multizone air handling unit drawing and answer the questions.

1. _____ Are receiver controllers 3 and 4 single- or dual-input receiver controllers?

2. _____ Receiver controller 3 setpoint is ___.

3. _____ Receiver controller 4 setpoint is ___.

4. _____ Receiver controller 3 action is ___.

5. _____ Receiver controller 4 action is ___.

6. _____ The branch line of receiver controller 3 is piped to ___.

7. _____ At a return air relative humidity of 30%, the output pressure of receiver controller 3 is ___ psig.

8. At a discharge air relative humidity of 85%, is receiver controller 3 able to open the humidifier valve? Why or why not?

_____

_____

9. _____ At an outside air temperature of 75°F, the pressure at port P of switching relay 1 is ___ psig.

Activity 10-4. Receiver Controller Troubleshooting

Scenario 1: A complaint is received that the air is too hot. The outside air temperature is 75°F. The mixed air dampers are wide open.

1. _____ The output pressure gauge on receiver controller 5 is 0 psig. Is this correct?

2. How should the operation of the controller be checked?

_____

_____

Scenario 2: A complaint is received that the air is too cold. The outside air temperature is 35°F. Inspection of the unit indicates that the hot duct temperature is approximately 85°F.

3. _____ Is this correct?

4. How should the operation of the controller be checked?

_____

_____

Scenario 3: A complaint is received that the humidity level is too low. The return air humidity is measured and found to be 30% rh and the discharge air humidity is 27% rh.

5. _____ The output pressure gauge on receiver controller 3 reads 20 psig. Is this correct?

6. _____ The output pressure gauge on receiver controller 4 reads 0 psig. Is this correct?

7. Which controller has a problem and how would it be corrected?

_____

_____

Name: _____ Date: _____

Control System Principles

_____ **1.** A(n) ___ is a device used in a control system that produces a desired function when actuated by the output signal from a controller.

      A. controller
      B. auxiliary device
      C. controlled device
      D. booster relay

_____ **2.** A(n) ___ is a device that allows two different types of components, voltage levels, voltage types, or systems to be interconnected.

T   F   **3.** A switching relay is a device that switches air flow from one circuit to another.

_____ **4.** Switching relays are used to create ___ circuits.

_____ **5.** A booster relay is a device that ___ the air volume available to a damper or valve while maintaining the air pressure at a 1:1 ratio.

      A. increases
      B. decreases
      C. maintains
      D. varies

_____ **6.** A ___ is a multiple-input device that selects the higher or lower of two pneumatic signal levels.

      A. controller
      B. transducer
      C. booster relay
      D. signal selection relay

_____ **7.** The branch line pressure of pneumatic direct-acting thermostats ___ as temperature increases.

T   F   **8.** A low signal selection relay is a relay that provides measurement of a large number of zones to ensure accurate zone signal measurement.

_____ **9.** A(n) ___ is a relay that prevents outside air dampers from completely closing.

_____ 10. A(n) ___ is an auxiliary device mounted to a damper or valve actuator that ensures that the damper or actuator moves to a given extension.

_____ 11. Biasing relays enable multiple heating and cooling devices to be placed in the ___.

    A. same cabinet
    B. correct sequence
    C. same system
    D. proper spring range

_____ 12. A(n) ___ is a device that allows an air pressure signal to energize or de-energize an electrical device such as a fan, pump, compressor, or electric heating device.

T   F   13. Pneumatic/electric switches perform a task that is the opposite of that of electric/pneumatic switches, and the switches cannot be interchanged.

_____ 14. A(n) ___ is a device that changes one type of proportional control signal into another.

_____ 15. An electronic/pneumatic transducer (EPT) is a device that converts a(n) ___ signal to a(n) ___ signal.

    A. analog electrical input; digital output
    B. digital input; analog electrical output
    C. electronic input; air pressure output
    D. air pressure input; electronic output

_____ 16. A(n) ___ is a device that protects against damage due to a low temperature condition.

_____ 17. Low temperature cutout control setpoints are adjustable, with the most common being between ___ and ___.

    A. 25°F, 32°F
    B. 30°F, 35°F
    C. 35°F, 40°F
    D. 60°F, 65°F

T   F   18. Thermostats in signal selection systems must be calibrated at a higher output pressure and have a lower setpoint.

_____ 19. Auxiliary devices that change pressure do not change the ___ of air flow.

_____ 20. A(n) ___ is a signal selection relay that selects the higher of two input pressures and outputs the higher pressure to a controlled device.

Switching Relays

_____ **1.** valve

_____ **2.** low-limit transmitter

_____ **3.** restrictor

_____ **4.** coil

_____ **5.** thermostat

_____ **6.** switching relay

_____ **7.** master air supply

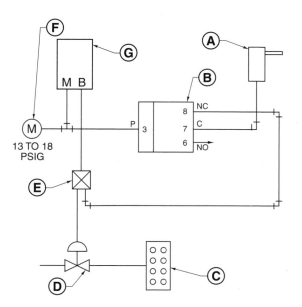

Low Signal Selection Relays

_____ **1.** exhaust

_____ **2.** direct-acting pneumatic thermostat

_____ **3.** low signal selection relay

_____ **4.** restrictor

_____ **5.** normally open heating valve

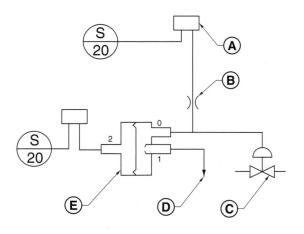

Minimum-Position Relays

_____   **1.** mixed-air plenum

_____   **2.** outside air lockout

_____   **3.** pneumatic low-limit thermostat

_____   **4.** cool outside air

_____   **5.** outside air damper

_____   **6.** averaging element

_____   **7.** minimum position relay

_____   **8.** warm return air

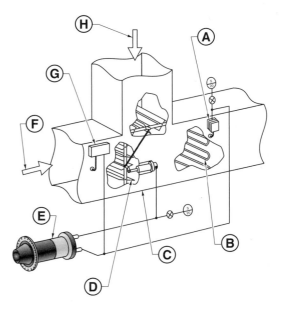

Electronic/Pneumatic Transducers

_____   **1.** wiring to electronic circuit

_____   **2.** tubing to pneumatic circuit

_____   **3.** pressure gauge

_____   **4.** electronic/pneumatic transducers

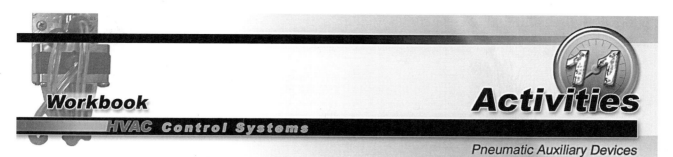

Name: _____ Date: _____

MULTIZONE AIR HANDLING UNIT

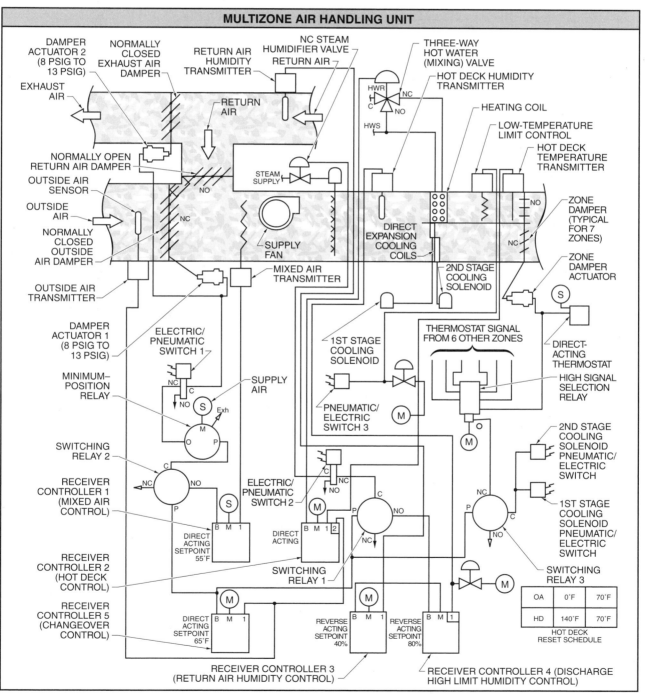

OA	0°F	70°F
HD	140°F	70°F

HOT DECK
RESET SCHEDULE

Activity 11-1. Switching Relays

Locate all switching relays on the multizone air handling unit drawing and answer the questions.

1. _____ How many switching relays are used in the multizone air handling unit?

2. List the switching relays.

_____

_____

_____

3. Which two ports are connected if the pilot pressure is 0 psig?

_____

_____

4. Which two ports are connected if the pilot pressure is 20 psig?

_____

_____

5. _____ Is the switch point adjustable?

6. Why is the NC port of switching relay 2 switched to atmosphere?

_____

_____

_____

7. What control causes switching relay 1 to switch? Under what condition(s)?

_____

_____

_____

Activity 11-2. Electric/Pneumatic Switches

Locate all electric/pneumatic switches on the multizone air handling unit drawing and answer the questions.

1. _____ How many electric/pneumatic switches are used in the multizone air handling unit?

2. List the electric/pneumatic switches.

_____

_____

_____

3. If the electric/pneumatic solenoid is energized, which two ports are connected?

_____

_____

4. If the electric/pneumatic solenoid is de-energized, which two ports are connected?

_____

_____

5. If electric/pneumatic switch 2 is de-energized, what happens to the air at the common port? What does this cause at the humidifier valve?

_____

_____

_____

6. If electric/pneumatic switch 1 is connected to the supply fan circuit, what happens at the outside air damper if the fan shuts down? Why would this be desirable?

_____

_____

_____

Activity 11-3. Pneumatic/Electric Switches

Locate all pneumatic/electric switches on the multizone air handling unit drawing and answer the questions.

1. _____ How many pneumatic/electric switches are used in the multizone air handling unit?

2. List the pneumatic/electric switches.

_____

_____

_____

3. Which electrical contacts are connected at 0 psig?

_____

_____

4. Which electrical contacts are connected at 20 psig?

_____

_____

5. _____ Do both pneumatic/electric switch 1 and pneumatic/electric switch 2 have the same setpoint?

6. If pneumatic/electric switch 1 controls the first stage cooling at one-third capacity and pneumatic/electric switch 2 controls the second stage cooling, estimate their setpoints.

_____

_____

_____

7. Pneumatic/electric switch 3 is piped to the low-temperature limit control (thermostat). If the low-temperature limit thermostat bleeds the air off of the pneumatic/electric switch, what happens to the supply fan?

_____

_____

_____

Activity 11-4. High Signal Selection Relay

Locate the high signal selection relay on the multizone air handling unit drawing and answer the questions.

1. What condition does the highest pressure value represent?

_____

_____

2. _____ Should the zone thermostats be calibrated the same?

Activity 11-5. Auxiliary Device Troubleshooting

Scenario 1: A complaint that rooms are too hot has been received. The outside air temperature is 75°F. The highest zone pressure is 20 psig. Only stage 1 of cooling is running. A voltmeter placed across NO and C terminals of the second stage pneumatic/electric switch reads 110 VAC.

1. _____ Is the switch open or closed?

2. What possible problem might this indicate?

_____

_____

_____

Scenario 2: A complaint that rooms are too dry has been received. The outside air temperature is 30°F. The humidifier valve is closed. The humidity controller is calling for full humidity. There is 0 psig on the humidifier valve. An ohmmeter placed across the solenoid of electric/pneumatic switch 2 reads infinite resistance.

3. What is a possible problem with the solenoid?

_____

_____

_____

Name: _____ Date: _____

Control System Principles

_____ 1. A(n) ___ is a pictorial and written representation of pneumatic controls and related equipment.

_____ 2. Pneumatic control diagrams identify proper ___ of equipment, as well as potential trouble spots and design flaws.

_____ 3. A(n) ___ is a drawing of a mechanical system that illustrates actual controls and piping between devices.

        A. parts list
        B. interface diagram
        C. control diagram
        D. sequence chart

T   F 4. A sequence chart is a reference list that indicates part description acronyms and actual manufacturer part names and numbers.

_____ 5. A written sequence of operation is a written description of the ___ of a control system.

_____ 6. A(n) ___ is a chart that shows the numerical relationship between the different values in a pneumatic system.

_____ 7. An electrical interface diagram is a drawing showing the interconnection between the ___ components and ___ equipment in a system.

        A. pneumatic; electronic
        B. pneumatic; electrical
        C. electronic; electrical
        D. digital; automated

_____ 8. ___ control is the most basic control method of any mechanical system.

_____ 9. The fan of a constant-volume air handing unit always runs at ___ of its rated capacity.

        A. 25%
        B. 50%
        C. 75%
        D. 100%

_____ **10.** Most variable air volume air handling units maintain a constant static pressure of ___ and a constant discharge air temperature of ___.

    A. 1″ wc; 32°F
    B. 1″ wc; 55°F
    C. 14.7 psia; 32°F
    D. 14.7 psia; 72°F

T   F   **11.** When excessively low temperatures are sensed in a variable air volume air handling unit after the cooling coil, the low-limit electric thermostat de-energizes the supply fan.

_____ **12.** In pressure-dependent variable air volume systems, a thermostat controls the damper actuator directly without reference to the ___ of air flowing through the ductwork.

    A. volume
    B. pressure
    C. temperature
    D. humidity

_____ **13.** In pressure-independent variable air volume systems, a(n) ___ measures air flow through a variable air volume terminal box.

_____ **14.** A(n) ___ is a small air handling unit mounted on the outside wall of each room of a building.

_____ **15.** A(n) ___ is an evaporative water cooler that uses natural evaporation to cool water.

    A. air chiller
    B. liquid chiller
    C. heat exchanger
    D. cooling tower

_____ **16.** A(n) ___ is a heat exchanger that removes heat from high-pressure refrigerant vapor.

T   F   **17.** A heater may be included in a cooling tower sump to keep the water temperature above freezing for winter operation.

_____ **18.** Multizone air handling units are identified by the number of ___ mounted on the discharge end of the air handler.

_____ **19.** Small package boilers have integral electric control systems, but ___ are often used for maintaining the water loop temperature, including the starting and stopping of system pumps.

_____ **20.** When zone temperature is satisfied by a variable volume terminal box, the ___ controller maintains a minimum volume of air flow through the terminal box for ventilation.

Single-Zone Air Handling Unit

_____ 1. to fan circuit

_____ 2. mechanical interlock

_____ 3. supply fan

_____ 4. zone thermostat

_____ 5. main air supply

_____ 6. cold water valve

_____ 7. electric/pneumatic switch

_____ 8. filter

_____ 9. supply air

_____ 10. exhaust air

_____ 11. damper actuator

_____ 12. return air

_____ 13. normally open hot water valve

_____ 14. outside air

_____ 15. minimum position relay

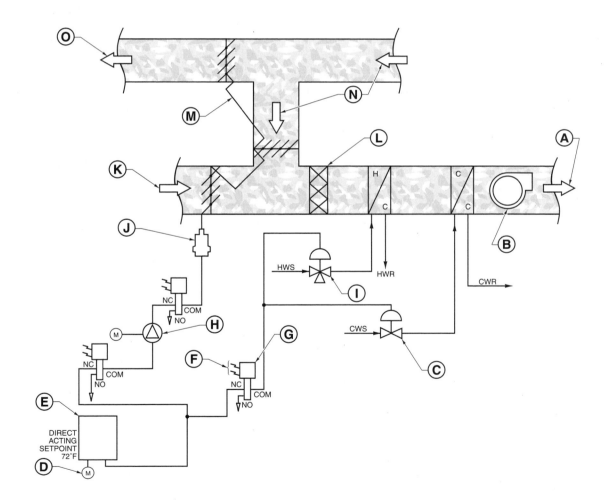

Variable Air Volume Air Handling Unit

_____ 1. switching relay

_____ 2. supply fan

_____ 3. damper actuator

_____ 4. duct static pressure transmitter

_____ 5. cooling coil

_____ 6. outside air damper

_____ 7. discharge air temperature transmitter

_____ 8. minimum position relay

_____ 9. fan static control receiver controller

_____ 10. three-way chilled water (mixing) valve

_____ 11. mixed air temperature transmitter

_____ 12. low-limit electric thermostat

_____ 13. filter

_____ 14. mixed air control receiver controller

_____ 15. electric/pneumatic switch

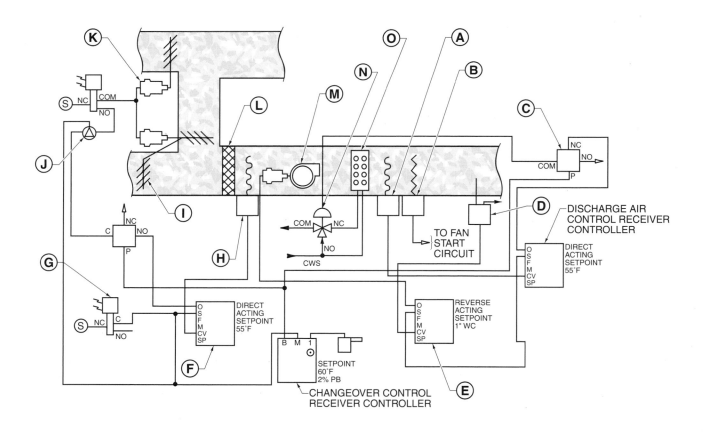

Boiler Control

_____ 1. zone hot water valve

_____ 2. boiler

_____ 3. hot water return transmitter

_____ 4. changeover control receiver controller

_____ 5. heating/cooling coil

_____ 6. three-way hot water valve

_____ 7. restrictor

_____ 8. circulating pump

_____ 9. zone thermostat

_____ 10. outside air transmitter

_____ 11. pneumatic/electric switch

_____ 12. hot water control receiver controller

_____ 13. panel receiver gauge

_____ 14. hot water supply transmitter

_____ 15. pump status differential pressure switch

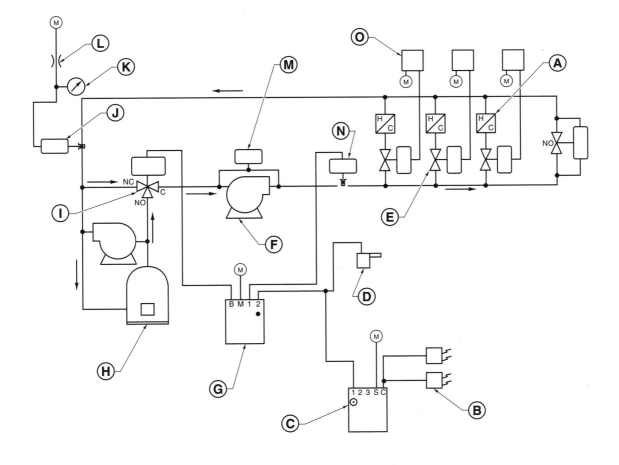

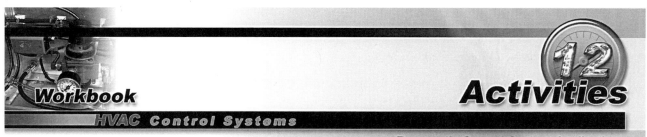

Name: _____ Date: _____

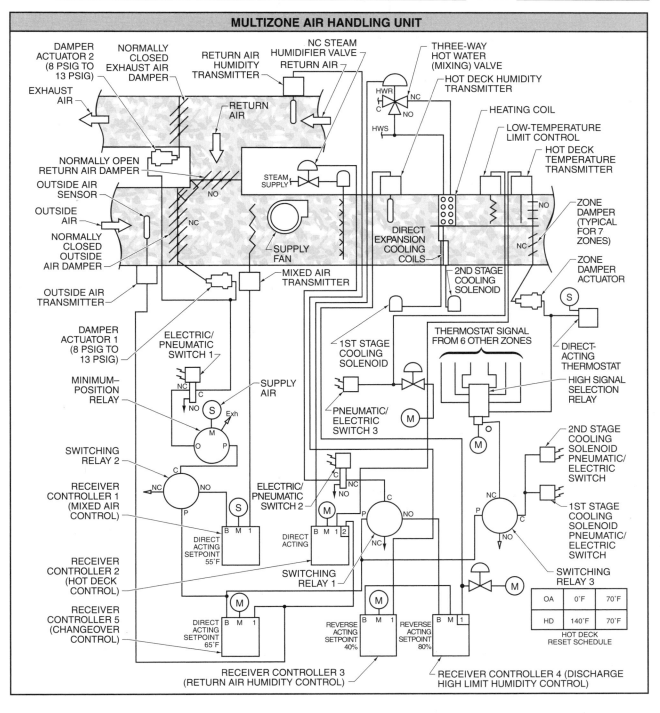

MULTIZONE AIR HANDLING UNIT

Activity 12-1. Air Handling Unit Heating Control System

Use the multizone air handling unit drawing to answer the questions. Trace the heating controls of the system. The heating system controls include receiver controller 2 and the three-way hot water valve.

1. _____ Are there any electric/pneumatic switches or switching relays between the output of receiver controller 2 and the three-way hot water valve?

2. _____ If so, does their operation affect receiver controller 2 or the valve?

3. _____ Which controller shares the outside air temperature sensor with the reset transmitter port of receiver controller 2?

4. _____ If the outside air transmitter fails, does it affect both controllers?

5. _____ Is the operation of the hot deck controller tied into the supply fan?

Activity 12-2. Air Handling Unit Cooling Control System

Use the multizone air handling unit drawing to answer the questions. Trace the cooling controls of the system. The cooling system controls include receiver controller 5, switching relay 3, first and second stage solenoid pneumatic/electric switches, high signal selection relay, first and second stage cooling solenoids, and zone thermostat.

1. _____ If receiver controller 5 (changeover control) is operating properly, its branch pressure when the temperature is above 65°F is ___ psig.

2. _____ At a temperature above 65°F, what two ports of switching relay 3 are connected?

3. _____ Does the highest pressure from the signal selection relay then control the pneumatic/electric switches?

4. _____ If receiver controller 5 (changeover control) is operating properly, its branch pressure is ___ psig when the temperature is below 65°F.

5. _____ At a temperature below 65°F, what two ports of switching relay 3 are connected?

6. _____ Does the highest pressure from the signal selection relay then control the pneumatic/electric switches?

7. _____ The room with the highest demand for cooling controls the mechanical cooling equipment. If a room is calling for maximum cooling, would the damper for that zone be fully open to provide cool air to that zone?

8. _____ Would the setpoints of the first and second stage pneumatic/electric switches be coordinated with the spring range of the damper actuator?

9. If this is true, and the zone damper spring range is 3 psig to 9 psig, give sample setpoints for the cooling solenoid pneumatic/electric switches.

Activity 12-3. Air Handling Unit Damper Control System

Use the multizone air handling unit drawing to answer the questions. Trace the damper controls of the system. The damper controls include receiver controller 1, receiver controller 5, switching relay 2, the minimum-position relay, electric/pneumatic switch 1, and damper actuator 1 and 2.

1. _____ The setpoint of receiver controller 1 (mixed air control) is ___°F.

2. _____ If receiver controller 5 (changeover control) is operating properly, its branch pressure is ___ psig when the temperature is above 65°F.

3. _____ At a temperature above 65°F, what two ports of switching relay 2 are connected?

4. At a temperature above 65°F, what happens to the air at the pilot port of the minimum position relay?

_____

_____

5. _____ At a temperature above 65°F, does the minimum position relay allow the outside air damper to fully close?

6. _____ For a given 8 psig to 13 psig spring range, what is a possible minimum position relay setpoint for 10% minimum outside air?

7. If electric/pneumatic switch 1 is tied into the fan operation, what happens to the outside air damper if the fan shuts down?

_____

_____

8. Why might this be desired?

_____

_____

9. _____ If receiver controller 5 (changeover control) is operating properly, its branch pressure is ___ psig at a temperature below 65°F.

10. _____ At a temperature below 65°F, what two ports of switching relay 3 are connected?

11. _____ Does receiver controller 1 then control the outside air, return air, and exhaust air dampers to maintain its setpoint?

Activity 12-4. Air Handling Unit Humidity Control System

Use the multizone air handling unit drawing to answer the questions. Trace the humidity controls of the system. The humidity controls include receiver controller 3, receiver controller 4, receiver controller 5, switching relay 1, electric/pneumatic switch 2, and the normally closed steam humidifier valve.

1. _____ If receiver controller 5 (changeover control) is operating properly, its branch pressure is ___ psig when the temperature is above 65°F.

2. _____ At a temperature above 65°F, what two ports of switching relay 1 are connected?

3. _____ Can the humidifier valve open when the outside air temperature is above 65°F?

4. _____ If electric/pneumatic switch 2 is connected to the operation of the supply fan, will the humidifier operate if the supply fan is off?

5. Why would this be desired?

_____

_____

6. Can mechanical cooling and the humidifier valve operate at the same time? Why or why not?

_____

_____

7. High receiver controller 4 (discharge high limit humidity control) has a setpoint of 80% rh. If the discharge humidity reaches a level above 80%, what does receiver controller 4 do to the steam humidifier valve?

_____

_____

8. Why would this be desired?

_____

_____

Activity 12-5. Air Handling Unit Low Temperature Limit Control System

Use the multizone air handling unit drawing to answer the questions. Trace the low temperature limit controls of the system. The low temperature limit controls include the low temperature limit control and pneumatic/electric switch 3.

1. _____ If it is desired to prevent a freezing condition after the heating coil, the low temperature limit would trip at ___°F.

2. When pneumatic/electric switch 3 stops the operation of the supply fan, what happens to the humidifier?

_____

3. When pneumatic/electric switch 3 stops the operation of the supply fan, what happens to the outside air dampers?

_____

4. _____ Is the operation of the hot deck controller tied into the low-temperature limit control?

Activity 12-6. Air Handling Unit Sequence of Operation

Use the single-zone air handling unit drawing to answer the questions. Circle the heating valve, cooling valve, and damper operation for your reference.

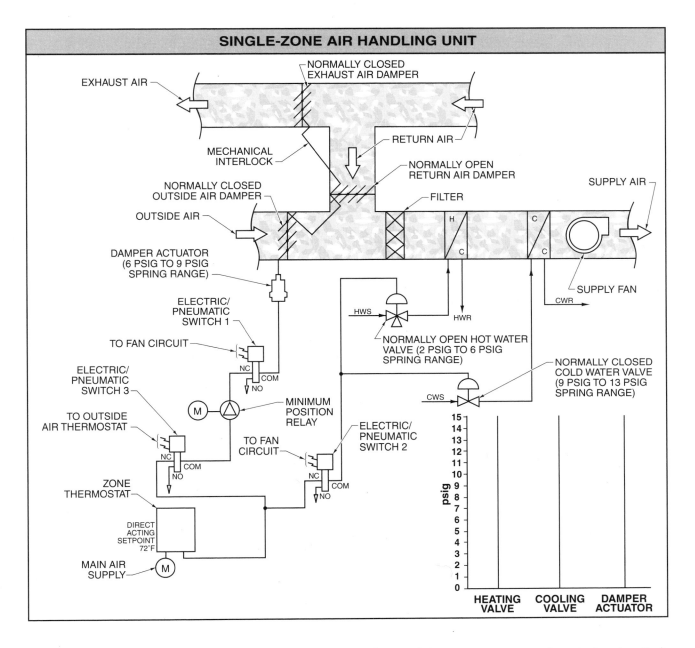

1. Mark the operation of the heating valve, cooling valve, and damper actuator on the graph using their indicated spring ranges.

2. _____ Is there any overlap between heating, cooling, and damper operations?

3. _____ If spring range shift occurs for any of the devices, could there be simultaneous heating, cooling, or damper operation?

4. Write the heating sequence of operation for the single-zone air handling unit.

_____

_____

_____

_____

_____

5. Write the cooling sequence of operation for the single-zone air handling unit.

_____

_____

_____

_____

_____

6. Write the damper sequence of operation for the single-zone air handling unit.

_____

_____

_____

_____

Name: _____ Date: _____

Control System Principles

T   F   **1.** An electrical control system is a control system that uses electricity (24 VAC or higher) in combination with a mechanism such as a pivot, mechanical bellows, or other device.

_____ **2.** Electricity is the energy released by the flow of ___ in a conductor (wire).

        A. neutrons
        B. protons
        C. electrons
        D. nucleuses

_____ **3.** ___ is the amount of electrons flowing through a conductor.

        A. Resistance
        B. Voltage
        C. Current
        D. none of the above

_____ **4.** ___ is the international unit of frequency equal to one cycle per second.

_____ **5.** Electrical control systems at ___ are the most common line-voltage control systems.

_____ **6.** A ___ is a switch that isolates electrical circuits from the voltage source to allow safe access for maintenance or repair.

        A. relay
        B. DPDT switch
        C. pressure switch
        D. disconnect

_____ **7.** ___ is a device that shuts OFF the power supply when current flow is excessive.

T   F   **8.** A contactor is an electrically operated switch that includes motor overload protection.

_____ **9.** ___ is the increase of a controlled variable above the controller setpoint.

_____ **10.** A(n) ___ can be used to ensure proper air and water flow by measuring the differential pressure across a fan or pump.

_____ **11.** A(n) ___ is a mechanical procedure that consists of reversing refrigerant flow in a system to melt ice that builds up on the evaporator coil.

_____ **12.** A hygroscopic element is a device that changes its characteristics as ___ changes.

    A. humidity
    B. temperature
    C. current
    D. voltage

_____ **13.** In a refrigeration control application, when power is applied to the HVAC unit, the ___ is energized.

_____ **14.** Voltage is measured across an electrical control device by placing the DMM leads in ___ across the device being tested.

    A. loops
    B. combination
    C. series
    D. parallel

T   F   **15.** A clamp-on meter is a device designed to measure current in a circuit by measuring the strength of the magnetic field around a single conductor.

_____ **16.** ___ is the presence of a complete path for current flow.

_____ **17.** In HVAC, all applicable safety codes and regulations must be followed per the ___ and/or authority having jurisdiction.

_____ **18.** A(n) ___ is when current leaves the normal current-carrying path by going around the load and back to the power source or to ground.

    A. dead short
    B. short circuit
    C. partial short
    D. open circuit

_____ **19.** ___ is current that reverses its direction of flow at regular intervals.

_____ **20.** A step-down transformer is a transformer in which the secondary coil has ___ turns of wire than the primary coil.

Series Circuit Connections

_____ **1.** control relay coil

_____ **2.** pressure switch

_____ **3.** current flow

_____ **4.** one path for current flow

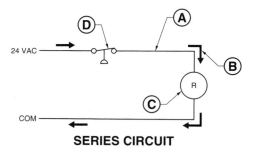

SERIES CIRCUIT

Contactors and Motor Starters

_____ **1.** 3ϕ power supply lines

_____ **2.** motor power wires

_____ **3.** normally closed auxiliary contact terminals

_____ **4.** coil terminals

_____ **5.** overload contact

_____ **6.** power supply terminals

_____ **7.** coil cover

_____ **8.** secondary terminals

_____ **9.** auxiliary contact

_____ **10.** heaters

_____ **11.** coil

_____ **12.** normally open auxiliary contact terminals

_____ **13.** auxiliary contact manual activation button

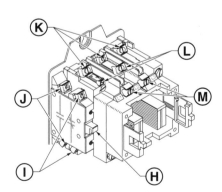

CONTACTOR

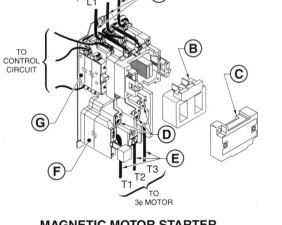

MAGNETIC MOTOR STARTER

Pressure Switches

_____ **1.** metal bellows

_____ **2.** neoprene diaphragm

_____ **3.** atmospheric pressure

_____ **4.** output to mechanical linkage and switch

_____ **5.** system pressure

_____ **6.** output force to linkage

_____ **7.** system pressure connection

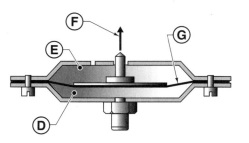

DIAPHRAGM ELEMENT

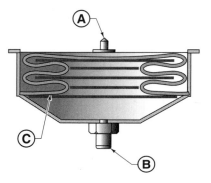

BELLOWS ELEMENT

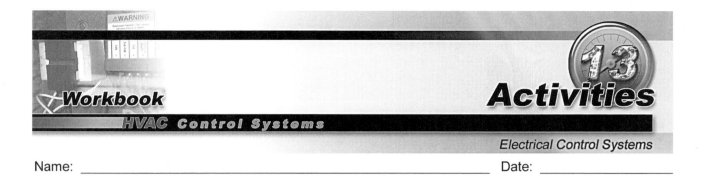

Name: _____ Date: _____

Activity 13-1. Relay Troubleshooting

It is summer. A too-hot complaint is received. Upon arrival in the room, it is determined that the indoor fan is running but the compressor and outdoor fan motor are not. The thermostat fan switch is in the auto position. A digital multimeter (DMM) set to measure voltage is used to check the voltage in the circuit.

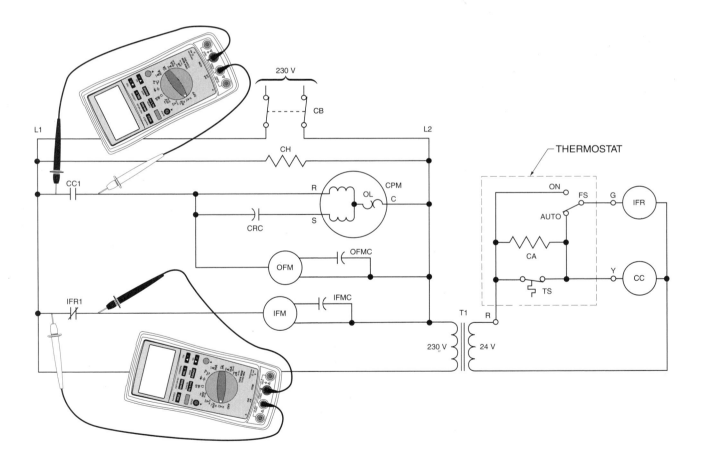

1. _____ With the indoor fan operating normally, the voltage across IFR1 is ___ VAC.

2. _____ With the compressor and outdoor fan motor not operating, the voltage across CC1 is ___ VAC.

Activity 13-2. Transformer Secondary Troubleshooting

It is summer. A too-hot complaint is received. Upon arrival in the room, it is determined that the indoor fan, compressor, and outdoor fan motor are not running. The thermostat fan switch is in the auto position. A DMM set to measure voltage is used to check the voltage across the circuit breaker. The voltage level is okay. The control transformer primary is checked and the voltage level is 230 VAC. The voltage across the control transformer secondary coil is also checked.

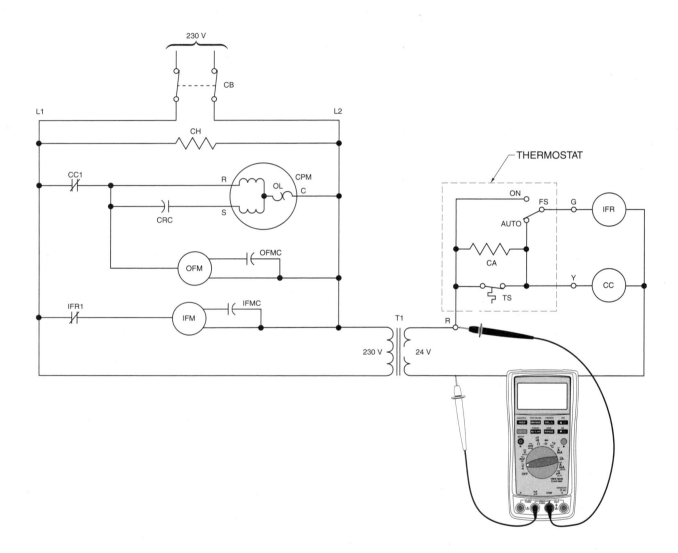

1. _____ The voltage across the control transformer secondary coil when it is open is ___ VAC.

Activity 13-3. Thermostat Troubleshooting

It is summer. A too-cold complaint is received. Upon arrival in the room, it is determined that the indoor fan, compressor, and outdoor fan motor are running. The thermostat is set at 74°F, but it is 68°F in the room. The thermostat fan switch is in the auto position. Regardless of the temperature at which the thermostat is set, the unit never shuts off. A bad thermostat is suspected. After locking and tagging out the power supply, a DMM set to measure resistance is used to check the thermostat.

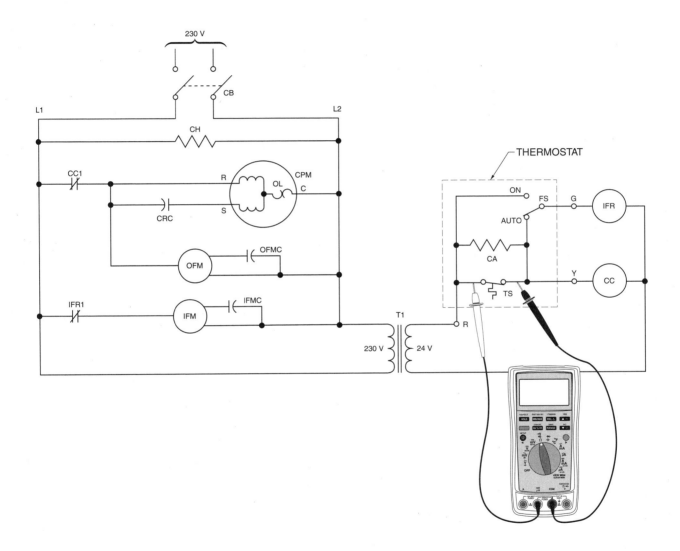

1. _____ The resistance value of the thermostat if its contacts are closed is ___ Ω.

Activity 13-4. Circuit Current Troubleshooting

It is summer. A too-hot complaint is received. The only diagnostic tool available is a digital clamp-on ammeter. Upon arrival, it is determined that the indoor fan is running but the compressor and outdoor fan motor are not. The full-load current of the indoor fan is 3 A, the compressor is 20 A, and the outdoor fan is 2 A. The thermostat fan switch is in the auto position. The jaws of the ammeter are clamped around the conductors in the circuit.

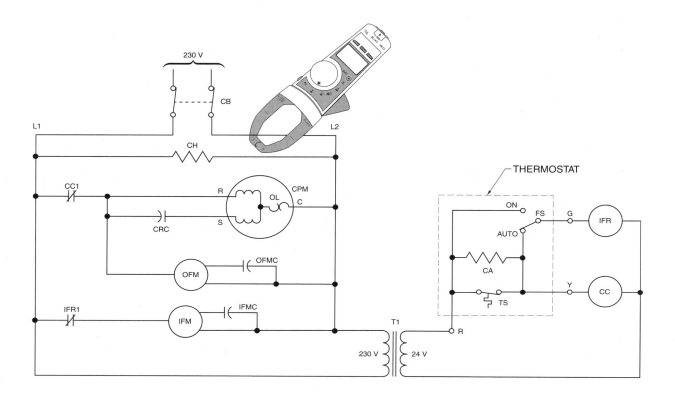

1. _____ The current reading of the clamp-on ammeter based on the these conditions is ___ A.

It is discovered that the thermostat was set to 80°F. After adjusting the thermostat to 74°F, the compressor and outdoor fan energize.

2. _____ The current reading of the clamp-on ammeter based on the new conditions is ___ A.

Name: _____ Date: _____

Control System Principles

_____ **1.** Electronic control systems use ___ electronic components to control HVAC equipment.

_____ **2.** ___ allow electrons to flow to plates that are located inside a glass housing.

    A. Vacuum tubes
    B. Transistors
    C. Rectifiers
    D. Light emitting diodes

_____ **3.** A(n) ___ is a material in which electrical conductivity is between that of a conductor (high conductivity) and that of an insulator (low conductivity).

T   F   **4.** Doping is the addition of material to a base element to alter the crystal structure of the element.

_____ **5.** A(n) ___ is a semiconductor device that allows current to flow in one direction only.

_____ **6.** A rectifier is a device that changes ___ voltage into ___ voltage.

    A. single-phase; three-phase
    B. three-phase; single-phase
    C. AC; DC
    D. DC; AC

_____ **7.** A(n) ___ is a circuit containing two diodes and a center-tapped transformer that permits both halves of the input AC sine wave to pass.

_____ **8.** A bridge rectifier is a circuit containing ___ diodes that permit both halves of the AC sine wave to pass.

    A. one
    B. two
    C. three
    D. four

_____ **9.** A(n) ___ is commonly used as a voltage shunt or electronic safety valve.

T   F   **10.** A light emitting diode is an electronic device that changes resistance or switches ON when exposed to light.

_____ **11.** A(n) ___ is a three-terminal semiconductor device that controls current flow according to the amount of voltage applied to the base.

_____ **12.** A bipolar junction transistor has its collector connected to the ___.

    A. load

    B. low power that functions as a valve

    C. common contact for the base and emitter circuits

    D. none of the above

_____ **13.** A(n) ___ is a transistor that consists of a phototransistor and a standard NPN transistor in a single package.

_____ **14.** A(n) ___ acts as a solid-state relay that can switch devices such as heaters, compressors, motors, and relays ON and OFF.

T   F   **15.** A silicon-controlled rectifier (SCR) is a thyristor that is capable of switching only direct current.

_____ **16.** A triac is a solid-state device used to switch ___.

    A. direct current

    B. alternating current

    C. devices with small resistance

    D. devices with high resistance

_____ **17.** A(n) ___ is an electronic device in which all components (transistors, diodes, and resistors) are contained in a single package or chip.

_____ **18.** An inexpensive ___ worn by technicians and connected to ground is used to protect against static electrical discharge to electronic components.

    A. glove

    B. wrist strap

    C. trace

    D. semiconductor device

_____ **19.** Temperature sensors use semiconductor materials that change ___ characteristics as the temperature around the sensor changes.

_____ **20.** Pressure sensors use a piezoelectric crystal, which changes ___ as pressure is exerted on the crystal.

Photodiodes

_____ **1.** lead

_____ **2.** metal can

_____ **3.** photodiode chip

_____ **4.** window

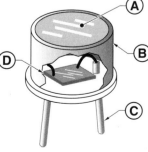

Metal-Oxide Semiconductor Field-Effect Transistor Construction

_____ **1.** drain

_____ **2.** P-type material

_____ **3.** gate

_____ **4.** metal-oxide material

_____ **5.** source

_____ **6.** insulator

_____ **7.** N-type material

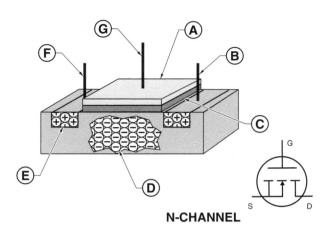

N-CHANNEL

Light Emitting Diodes

_____ **1.** anode lead

_____ **2.** epoxy housing

_____ **3.** flat mark

_____ **4.** cathode lead

_____ **5.** LED chip

_____ **6.** reflector

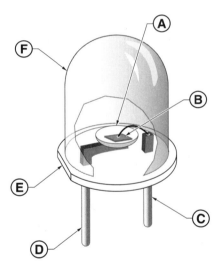

Bipolar Junction Transistor Construction

_____ **1.** emitter

_____ **2.** N-type material

_____ **3.** base

_____ **4.** P-type material

_____ **5.** collector

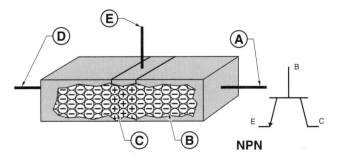

NPN

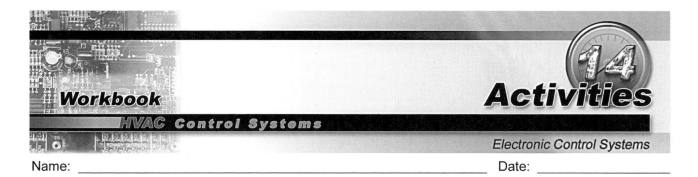

Name: _____ Date: _____

Activity 14-1. Operator Interfaces

A too-hot complaint is received from one of the rooms in a building. The LED display on a building automation system operator interface is used to determine which temperature sensor is in alarm and its value. Use the operator interface information to answer the questions.

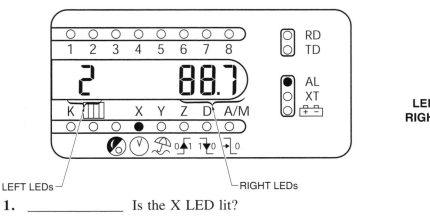

X = Analog Input Mode
AL = Point Is in Alarm
LEFT LEDs = Item Number
RIGHT LEDs = Value

1. _____ Is the X LED lit?

2. What does this indicate?

_____

3. _____ What is the value of the left LEDs?

4. _____ The left LEDs indicate input number ___.

5. _____ The right LEDs indicate the number ___.

6. _____ If this is a temperature sensor, its value is ___.

7. _____ Is the AL LED lit?

8. What does this mean?

_____

_____

Use the operator interface information to answer the questions.

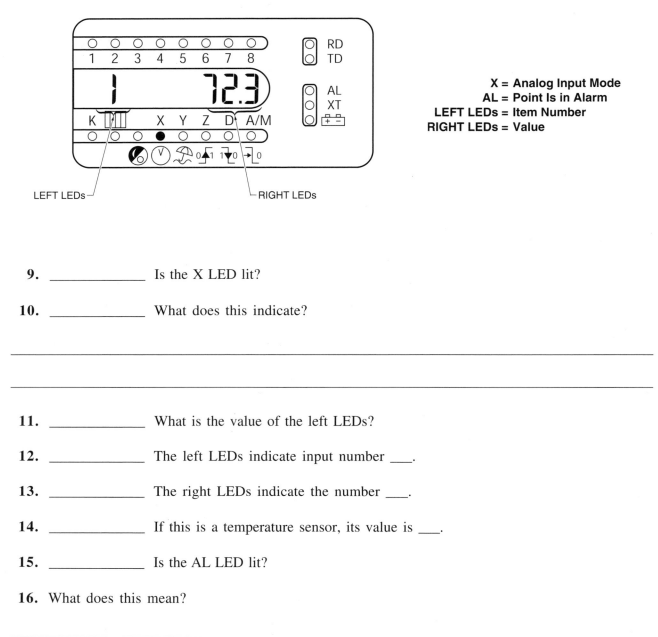

9. _____ Is the X LED lit?

10. _____ What does this indicate?

_____

_____

11. _____ What is the value of the left LEDs?

12. _____ The left LEDs indicate input number ___.

13. _____ The right LEDs indicate the number ___.

14. _____ If this is a temperature sensor, its value is ___.

15. _____ Is the AL LED lit?

16. What does this mean?

_____

_____

Activity 14-2. Electronic Circuit Troubleshooting

A too-cold complaint is received for one of the rooms in a building. Upon arrival in the mechanical room, it is determined that the air handling unit fan is off. Use the print to answer the questions.

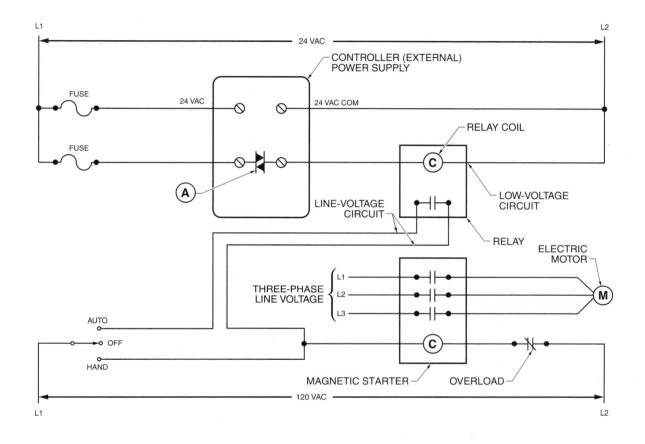

1. _____ Is the device labeled A a thyristor?

2. _____ What type of device is it?

3. _____ Is the device in an AC or DC circuit?

4. _____ The voltage level of the circuit is ___.

5. _____ What device interfaces the low-voltage circuit to the line-voltage circuit?

Name: _____ Date: _____

Control System Principles

_____ 1. A(n) ___ is a system that uses microprocessors (computer chips) to control the energy-using devices in a building.

_____ 2. A central supervisory control system is a control system in which the decision making equipment is located in ___ and the system enables/disables local (primary) controllers.

    A. one place
    B. various locations
    C. the air handling unit
    D. the building space controller

_____ 3. Some manufacturers provide ___ that allow central supervisory control systems to communicate with modern building automation system networks.

    A. unitary controllers
    B. universal input-output controllers
    C. gateway interface modules
    D. network communication modules

T   F 4. A central-direct digital control system is a control system in which all decisions are made in one location and that provides closed loop control.

_____ 5. Central-direct digital control systems control loads such as dampers, valves, and compressors through ___, without the need for local (primary) controllers.

_____ 6. A(n) ___ is an electronic device that follows commands sent to the device from the CPU of a central-direct digital control system.

_____ 7. Central-direct digital control systems are typically used for ___ instead of closed loop control.

    A. load control
    B. module control
    C. supervisory control
    D. sensor control

_____ 8. ___ is the recording of information such as temperature and equipment ON/OFF status at regular intervals.

_____ 9. ___ is information that is transmitted sequentially one bit (0 or 1) at a time.

_____ **10.** A(n) ___ is a notification of improper temperature or other condition existing in a building.

    A. setpoint
    B. network communication point
    C. communication bus
    D. alarm

_____ **11.** A(n) ___ is a control system that has multiple CPUs at the controller level.

T   F   **12.** Global data is data needed by all HVAC controllers in a network and includes outside air temperature and electrical demand.

_____ **13.** A(n) ___ is a controller designed to control only one type of HVAC system.

    A. application-specific controller (ASC)
    B. unitary controller
    C. universal input-output controller
    D. network communication controller

_____ **14.** A(n) ___ is an integrated-circuit memory chip that has an internal switch to permit the user to erase the memory contents and write new contents by means of electrical signals.

_____ **15.** Software is the ___ that enables a controller to function.

_____ **16.** A(n) ___ is a controller designed for basic zone control using a standard wall-mount temperature sensor.

    A. variable air volume box controller
    B. unitary controller
    C. universal input-output controller
    D. field interface controller

T   F   **17.** An air handling unit (AHU) controller is a controller that contains inputs and outputs required to operate large central-station air handling units.

_____ **18.** A(n) ___ is a controller that modulates the damper inside a variable air volume (VAV) terminal box to maintain a specific building space temperature.

_____ **19.** A universal input-output controller (UIOC) is a controller designed to control ___ HVAC equipment.

    A. one type of
    B. one piece of
    C. two pieces of
    D. most

_____ **20.** A network communication module (NCM) is a controller that coordinates communication from ___ to ___ on a network and provides a location for an operator interface.

    A. building automation system; operator interface
    B. gateway interface module; output device
    C. controller; output device
    D. controller; controller

Unitary Controller Rooftop Unit

_____ **1.** room temperature sensor

_____ **2.** exhaust air

_____ **3.** unitary controller

_____ **4.** heating and cooling stages

_____ **5.** return air

_____ **6.** outside air temperature sensor

_____ **7.** supply fan

_____ **8.** airflow switch

_____ **9.** economizer damper actuator

_____ **10.** supply air

_____ **11.** outside air

_____ **12.** supply air temperature sensor

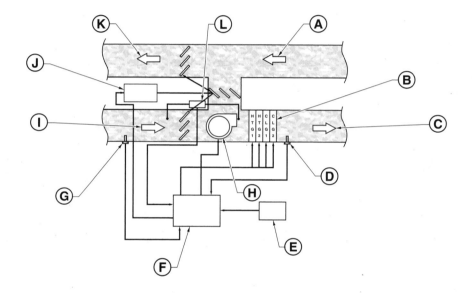

Pressure-Independent VAV Terminal Box with Reheat Coil

_____ **1.** primary air (55°F)

_____ **2.** room diffuser

_____ **3.** to controller

_____ **4.** reheat coil

_____ **5.** flow sensor at terminal box inlet

_____ **6.** VAV terminal box

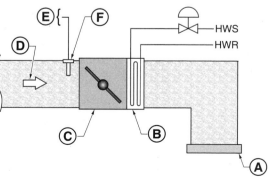

Air Handling Unit Controllers

_____ **1.** thermostat

_____ **2.** fan volume control

_____ **3.** filter

_____ **4.** static pressure sensor

_____ **5.** supply fan

_____ **6.** air handling unit controller

_____ **7.** return air

_____ **8.** mixed air plenum

_____ **9.** exhaust air

_____ **10.** volume damper

_____ **11.** cooling coil

_____ **12.** outside air

Workbook

HVAC Control Systems

Building Automation Systems and Controllers

Name: _____ Date: _____

Activity 15-1. Building Automation System Identification

Use the communication riser drawing to answer the questions.

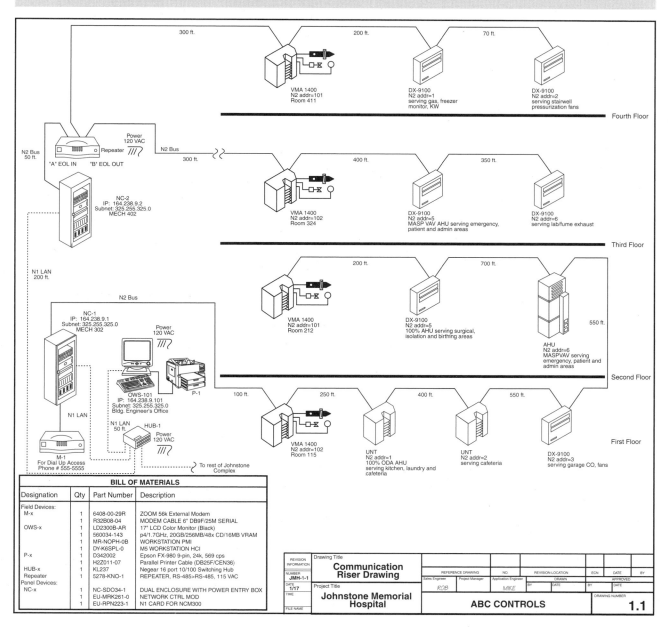

1. _____ Is the building automation system a central supervisory, central DDC, or distributed DDC system?

2. What items on the print might identify it as such?

_____

_____

3. _____ How many network communication modules are included in the system?

4. List each network communication module name and location.

_____

_____

5. What are the functions of the network communication module(s)?

_____

_____

6. _____ NC-2 is located a distance of ___′ from HUB-1.

7. _____ How many VAV terminal box controllers are included in the system?

8. List the location and local communication bus address for each VAV terminal box controller.

_____

_____

9. What type of programming would the VAV terminal box controllers use?

_____

_____

10. _____ In what kind of chip would their program be located?

11. _____ How many UIOMs are included in the system?

12. List the location, local communication bus address, and listed functions for each UIOM.

_____

_____

13. _____ Are the UIOMs easier or harder to program than the VAV terminal box controllers?

14. List the number, address, and HVAC unit served by the AHU controller(s).

_____

_____

> *An outside air temperature sensor is wired to the UIOM at address 3. The outside air temperature value must be shared with the AHU controller.*

15. What type of data is this and what controller is responsible for sharing the data?

_____

16. _____ The VAV terminal box controller in room 324 is located a distance of ___′ from the UIOM at address 5.

17. List the number, address, and function of all controllers located on the fourth floor.

_____

_____

18. _____ The hub part number is ___.

19. _____ The repeater part number is ___.

20. _____ The network control module part number is ___.

21. _____ The length of communication bus wire required for the job is ___′. *Note:* Add 10% for connections, wire stripping, etc.

Activity 15-2. Building Automation System Controller Specifications

> *Use the variable air volume box controller information on the following page to answer the questions.*

1. _____ What type of controller is it?

2. List the controller power requirements.

_____

3. List the controller primary control strategies.

_____

4. List the controller model numbers that have screw terminal block connections.

_____

5. Fan configurations that are available are ___.

6. List the number and type of analog inputs used with the AS-VAV1111-12 controller.

7. List the number and type of analog outputs used with the AS-VAV1111-12 controller.

8. _____ The required voltage and current of the binary outputs are ___.

Variable Air Volume Box Controller

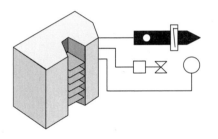

Description

The Variable Air Volume Box (VAV) Controller is specifically designed for digital control of single duct, dual duct, fan-powered, and supply/exhaust VAV box configurations. The controller can provide stand alone control of the VAV box, and can also integrate control of baseboard heat and lighting logic for the room or zone.

Features

- multiple modes of operation for various occupancy conditions
- stand alone control for small systems
- N2 bus communications and networking software capabilities
- interfaces to both electric and pneumatic actuators
- controller-resident performance calculations

OPTIONS	
Application	**Software**
Primary Equipment Types	VAX box, single duct, dual duct, fan powered or assisted, supply/exhaust
Primary Control Strategies	Pressure dependent, pressure independent, constant volume
Box Heat Configuration	Incremental, proportional, or two-position (NO or NC) valves 1-, 2-, or 3-stage electric
Baseboard Heat Configuration	Incremental, two-position, (NO or NC) valves, single stage electric
Cooling Configuration	Incremental output to damper actuator
Fan Configuration	Parallel, temperature setpoint; parallel, CFM setpoint; series, On-Off control; series, proportional control
Lighting Control	On and Off outputs to lighting relay in conjunction with occ/unocc mode
Unoccupied Control	Setup, setback, or shutdown

SPECIFICATIONS	
Variable Air Volume Box Controller	
Product	AS-VAV 1110-12, AS-VAV 1111-12, FA-VAV 1110-12, FA-VAV 1111-12 spade connector: AS-VAV 1140-12, AS-VAV 1141-12, FA-VAV 1140-12, FA-VAV 1141-12 screw terminal block
Ambient Operating Conditions	32˚F to 140˚F 10% to 90% RH
Dimensions (H x W x D)	6.5 x 6.4 x 4.0 in. 6.8 x 7.3 x 6.7 in. with enclosure
N2 Bus	Isolated
Zone Bus	8-pin phone jack or terminal block on controller
Power Requirements	24 VAC, 50/60 Hz, 10 VA plus binary output loads
Shipping Weight	2.6 lb

SELECTION CHART					
Model Number	**Termination Type**	**Analog Inputs**	**Binary Inputs**	**Analog Outputs**	**Binary Outputs**
AS-VAV1110-12 FA-VAV1110-12	Spade lug	6 RTD temperature element (NI, SI, or PT) 0-10 VDC transmitter 2 kΩ setpoint potentiometer	4 4-dry contacts 1 momentary pushbutton from zone sensor	—	8 24 VAC triacs at .5 A
AS-VAV1111-12 FA-VAV1111-12				2 0 VDC to 10 VDC at 10 mA	6 same as above
AS-VAV1140-12 FA-VAV1140-12	Screw terminal	6 RTD temperature element (NI, SI, or PT) 0-10 VDC transmitter 2 kΩ setpoint potentiometer	4 4-dry contacts 1 momentary pushbutton from zone sensor	—	8 24 VAC triacs at .5 A
AS-VAV1141-12 FA-VAV1141-12				2 0 VDC to 10 VDC at 10 mA	6 same as above

Name: _____ Date: _____

Control System Principles

_____ **1.** A(n) ___ is a device that allows an individual to access and respond to building automation system information.

    A. controller
    B. network communication module
    C. termination device
    D. operator interface

_____ **2.** ___ is a local area network (LAN) architecture that can connect up to 1024 nodes and supports data transfer rates of 10 megabits per second (Mbps).

_____ **3.** A local area network (LAN) is a communication network that spans a relatively ___ area.

    T   F **4.** A node is an output device or device actuator.

_____ **5.** The primary desktop PC used to communicate with a building automation system is commonly referred to as the ___.

    A. notebook
    B. front end
    C. dumb terminal
    D. main

_____ **6.** A(n) ___ is a printer used with a building automation system to produce hard copies of alarms (indications of improper system operation), preventive maintenance messages, and data trends.

_____ **7.** When loaded with the appropriate software, a(n) ___ can function as a portable service/maintenance tool that can be used to change setpoints and time schedules, or check a controller for proper operation.

_____ **8.** A(n) ___ is a small, lightweight, hand-held device that allows access to basic building automation system functions from various controllers throughout a building.

    A. dumb terminal
    B. portable operator terminal (POT)
    C. keypad display
    D. interface module

_____ 9. A(n) ___ is a controller-mounted device that consists of a small number of keys and a small display.

_____ 10. A dumb terminal can display ___ lines of information.

      A. 2 to 4
      B. 10 to 15
      C. 20 to 60
      D. 50 to 100

T   F   11. An off-site desktop PC uses a keypad interface module to dial in to a building automation system or allows the building automation system to dial in to an off-site desktop PC.

_____ 12. A(n) ___ is a base station for a notebook computer that includes a power supply, expansion slots, plus monitor and keyboard connections.

_____ 13. ___ is an automated phone service in which a voice on a computer chip prompts the caller to press various numbers for different functions when dialing with a standard touch-tone phone.

_____ 14. A building automation system can send alarms and logs to ___ in addition to off-site PCs.

      A. pagers
      B. fax machines
      C. cell phones
      D. all of the above

_____ 15. The operator interface for a building automation system is selected according to the ___ of interface required.

_____ 16. One of the primary reasons for poor customer satisfaction with a building automation system is the ___.

      A. lack of effective training
      B. complexity of the system
      C. cost of the system
      D. time to install

T   F   17. A comprehensive training plan should include building automation system training as well as HVAC system training and should be integrated with the human resources department of a facility.

_____ 18. A(n) ___ is the group of items and functions that an operator is permitted to perform.

_____ 19. The ___ allows the assigning of access levels and codes to other individuals.

_____ 20. Operators and technicians are issued ___ that activate the operator's or technician's access level.

Portable Operator Terminals

_____	**1.** alarm light
_____	**2.** display keys
_____	**3.** up/down arrow keys
_____	**4.** alarm triangle
_____	**5.** operating mode indicators
_____	**6.** on/off status indicators
_____	**7.** display item point list
_____	**8.** display areas
_____	**9.** display cursor dot

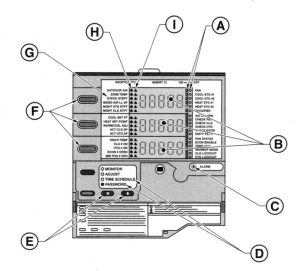

Off-Site Desktop PCs

_____	**1.** modem	_____	**4.** printer
_____	**2.** on-site desktop PC	_____	**5.** off-site desktop PC
_____	**3.** building automation system local area network	_____	**6.** network communication modules

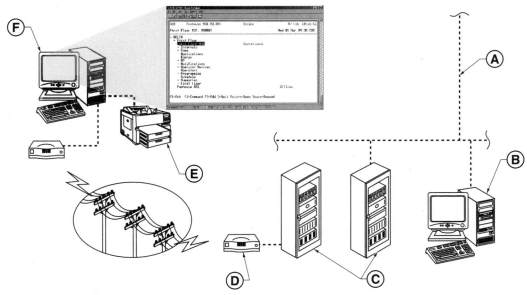

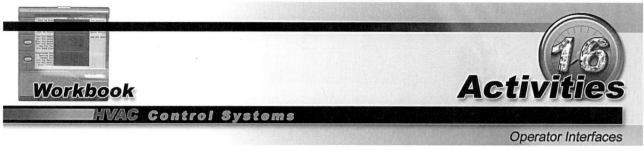

Name: _____ Date: _____

Activity 16-1. Operator Interface Identification

Use the communication riser drawing to answer the questions.

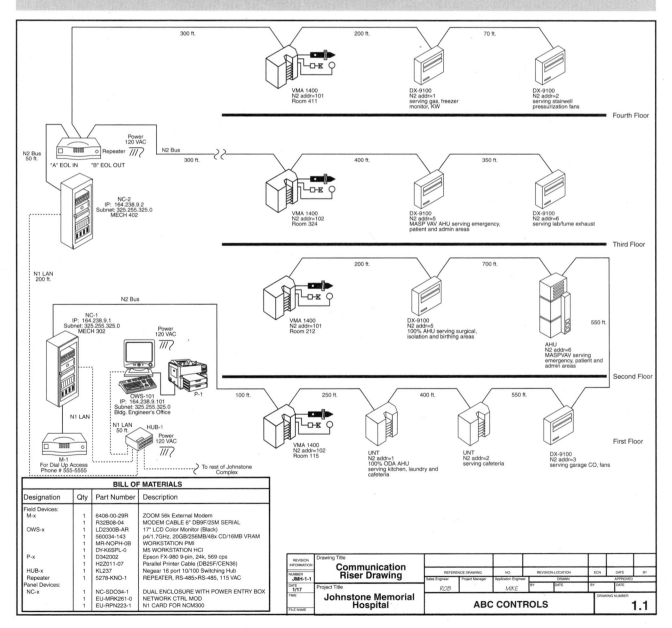

1. List three operator interface devices shown on the drawing.

   _____

   _____

2. _____ What type of interface device is OWS-101?

3. Where is OWS-101 located?

   _____

   _____

4. List three functions that OWS-101 might perform.

   _____

   _____

5. What is the purpose of interface device M-1?

   _____

   _____

6. _____ What is its phone number?

7. List three operator interface devices that M-1 might report to.

   _____

   _____

8. _____ What type of operator interface device is P-1?

9. _____ What operator interface device is it connected to?

10. List two functions that P-1 might perform.

    _____

    _____

Activity 16-2. Using Operator Interfaces

Use operator interface A to answer the questions.

1. _____ What type of operator interface device is shown?

2. List several advantages of this type of operator interface.

_____

_____

_____

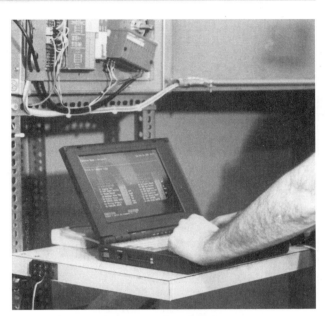

OPERATOR INTERFACE A

Use operator interface B to answer the questions.

3. _____ What type of operator interface device is shown?

4. In what types of situations is this type of operator interface best used?

_____

_____

OPERATOR INTERFACE B

5. List all devices on the communication riser diagram that would use the same type of operator interface.

_____

_____

Use the supervisory controller drawing to answer the questions.

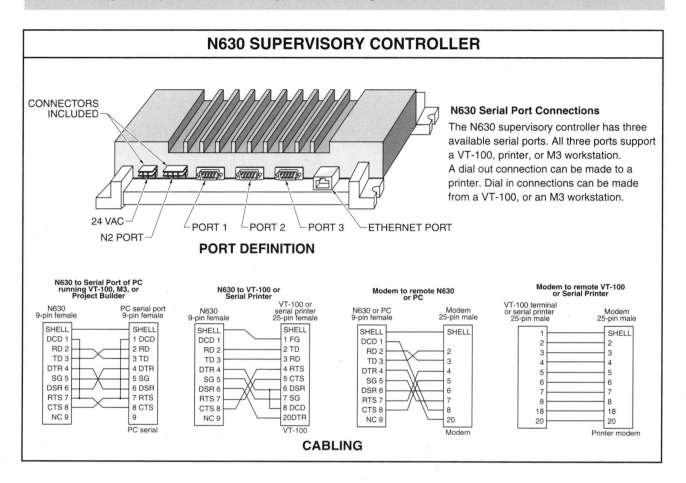

6. _____ How many 9 pin serial ports are shown?

7. In the cabling details, list three of the operator interface devices shown.

A security guard must be able to receive a list of alarms from the supervisory controller.

8. What are two interface devices that might be used?

This controller is being installed at a hospital. The system requires that changes be made from a home office.

9. List three operator interface devices that enable off-site changes.

Activity 16-3. Training and Access Levels

A building automation system installation contract calls for quality factory-certified training to be provided.

1. What can be done to determine what type of training is required?

_____

_____

Use the training schedule and the communication riser drawing to answer the question.

2. List three training courses that may be appropriate for the system being installed in the Johnstone Memorial Hospital.

_____

_____

There are eight staff members in the building. Joe is the supervisor. Sam and Susan are the lead technicians. John, Sarah, Jerry, Dave, and Jim are shift operators. List the appropriate access levels (supervisor, engineer, lead technician, and technician) needed for each.

3. _____ Joe access levels are ___.

4. _____ Sam access levels are ___.

5. _____ Susan access levels are ___.

6. _____ John access levels are ___.

7. _____ Sarah access levels are ___.

8. _____ Jerry access levels are ___.

9. _____ Dave access levels are ___.

10. _____ Jim access levels are ___.

ABC CONTROLS COMPANY - Training Schedule
Subject to change; courses will be on an as-needed basis.

Course #	Course Name	Course Start	Course End	Location of Training
120	Facility Operators	January 21	January 22	Anderson College
130	Controller Engineering/ Programming	February 2	February 8	Anderson College
121	ASC Engineering	February 18	February 22	Anderson College
120	Facility Operators	March 18	March 22	Anderson College
127	Hardware Troubleshooting	April 8	April 12	Anderson College
167	Graphics Development	April 23	May 1	Anderson College
100	HVAC Basics	May 3	May 11	Anderson College
134	VAV Operators	May 21	May 22	Anderson College
120	Facility Operators	June 1	June 7	Anderson College
164	Control Engineering	June 17	June 21	Anderson College
100	HVAC Basics	July 5	July 11	Anderson College
124	HVAC Maintenance	July 22	July 26	Anderson College
122	LAN Engineering	August 2	August 3	Anderson College
120	Facility Operators	September 3	September 11	Anderson College
132	Software Update	September 24	September 28	Anderson College
121	ASC Engineering	October 14	October 18	Anderson College
100	HVAC Basics	November 2	November 7	Anderson College
131	Graphics Engineering	November 18	November 22	Anderson College

To enroll, please call 1-800-555-4343.

Activity 16-4. Adding Operator Interfaces

A customer wants to purchase operator interface devices that are not part of the original job. Use the communication riser drawing to answer the questions.

1. _____ The system currently includes ___ (number) workstations.

2. _____ Can a new workstation be added in Mech Room 401?

3. _____ Can a printer be added?

4. _____ If a printer is added, can it be any printer?

5. _____ Can the network IP and subnet addresses be the same?

6. What suggestions can be made concerning suitable portable operator interface devices?

_____

_____

Name: _____ Date: _____

Control System Principles

T   F   **1.** A building automation system input is a device that senses and sends building condition information to a controller.

_____   **2.** A(n) ___ is a device that senses a variable such as temperature, pressure, or humidity and causes a proportional electrical signal change at the building automation system controller.

      A. analog input
      B. digital input
      C. analog output
      D. digital output

_____   **3.** A positive temperature-coefficient temperature sensor ___ its output resistance as the temperature increases, and ___ its output resistance as the temperature decreases.

      A. increases; increases
      B. increases; decreases
      C. decreases; decreases
      D. decreases; increases

_____   **4.** A(n) ___ is a resistor made of semiconductor material in which electrical resistance varies with changes in temperature.

_____   **5.** All manufacturers provide ___ to prevent direct sunlight from affecting outside air temperature sensor operation.

T   F   **6.** The most common humidity sensors measure percent relative humidity (% rh), while other sensors measure dewpoint or absolute humidity.

_____   **7.** Humidity sensors drift over time, with inaccuracies of ___ % per year being common.

      A. 1
      B. 2
      C. 3
      D. 5

_____   **8.** A(n) ___ is the amount of light produced by a lamp (lumens) divided by the area that is illuminated.

_____ 9. A ___ is a temperature-actuated switch.

    A. digital input
    B. timed override initiator
    C. transducer
    D. thermostat

_____ 10. A(n) ___ is a digital input device that switches open or closed because of the difference between two pressures.

_____ 11. A flow switch is a switch that contains a(n) ___ that moves when contacted by air or water flow.

_____ 12. A(n) ___ is a device that records the number of occurrences of a signal.

    A. current relay
    B. accumulator
    C. initiator
    D. none of the above

T   F 13. A current sensing relay is a device which surrounds a wire and detects the resistance due to electricity passing through the wire.

_____ 14. A timed override initiator is a device that, when ___, sends a signal to a controller which indicates that a timed override period is to begin.

_____ 15. A ___ is a device that changes the state of a controlled device in response to a command from a building automation system controller.

    A. valve or damper actuator
    B. controller
    C. triac
    D. building automation system output

_____ 16. A digital output is a device that accepts a(n) ___ signal.

_____ 17. A(n) ___ is a relay that requires a short pulse to energize the relay.

T   F 18. A pulse width modulated device is a digital output used to position a bi-directional electric motor.

_____ 19. Pulse width modulation (PWM) is a control technique in which a sequence of ___ is used to position an actuator.

    A. AC signals
    B. sine waves
    C. short pulses
    D. overvoltage spikes

_____ 20. A(n) ___ is a point that exists only in software and is not a hard-wired point.

Temperature Sensor Mounting

_____ **1.** well-mount

_____ **2.** duct-mount

_____ **3.** averaging

_____ **4.** wall-mount

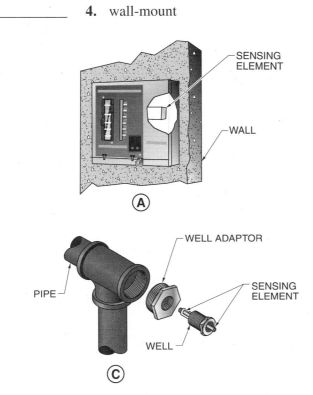

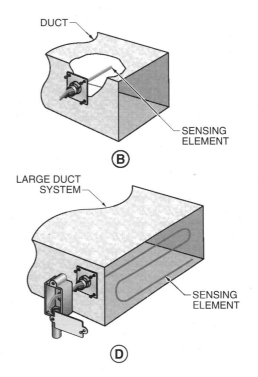

Humidity Sensor

_____ **1.** airflow

_____ **2.** output connection

_____ **3.** hygroscopic element

_____ **4.** duct

_____ **5.** strain gauge (to detect movement)

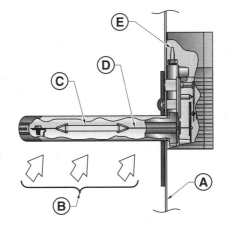

Workbook

HVAC Control Systems

Building Automation System Inputs and Outputs

Name: _____ Date: _____

Activity 17-1. Analog Inputs

Use the panel board wiring diagram to answer the questions.

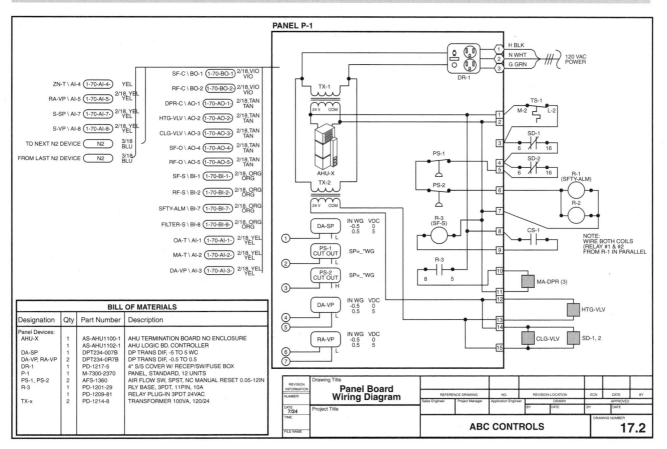

1. _____ How many analog inputs (AIs) are used in the system?

2. List three analog inputs.

_____

3. _____ How many temperature sensors are used in the system?

4. _____ The pressure range of the DA-SP sensor is ___″ wc.

5. _____ Should the supply air velocity pressure sensor be located as close as possible to an elbow or tee to obtain a better reading?

Use drawing details B20 and B22 to answer the questions.

6. _____ What color wire(s) is/are used to connect the field device to the AHU controller?

7. What other wires that are needed for the sensor are indicated on detail B20?

_____

_____

8. _____ What type of analog input is indicated on detail B22?

9. Must the controller programming be consistent with the type of sensor used in the application? Why or why not?

_____

_____

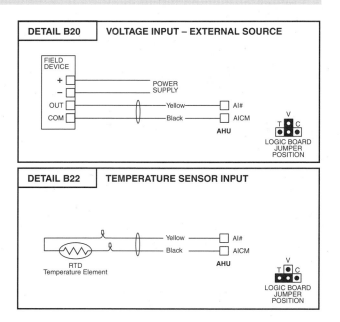

Use the temperature sensor/assembly information to answer the questions.

Temperature/Sensor Assembly

Description

TEM-161 completed assemblies are used in a variety of temperature sensing applications. Various other sensing elements and hardware configurations are available that can be field-assembled, depending on the application.

Applications

- control or indication of high-temperature steam using well-insertion assemblies in hot water pipes or tanks
- temperature-averaging
- duct-insertion for controlling cycling in areas of sudden, large temperature changes

SELECTION CHART		
Code Number	**Type**	**Description**
TEM-161-1	Nickel	17-ft averaging, temperature sensing element (1000 Ω, 1%) with handi-box
TEM-161-2	Nickel	Same as TEM-161-1 except that it has 8 ft averaging element
TEM-161-3	Nickel	High-temperature (550˚F) well-insertion element (1000 Ω, 1%)
TEM-161-4	Nickel	Same as TEM-161-2 except with setpoint, without cover
TEM-161-54	Silicon	Base room thermostat w/setpoint, wo/cover
TEM-161-55	Silicon	Space temperature assembly with wall plate adaptor and mounting bracket

SPECIFICATIONS		
TEM-161 Series Sensor/Hardware Assemblies		
Elements	TEM-161-1 through -8	Nickel wire resistance type
	TEM-161-54, -55	PTC Silicon
Reference Resistances	TEM-161-1 through -8	1000 Ω at 70˚F
	TEM-161-10	1000 Ω at 70˚F, 50% RH
	TEM-161-54, -55	1035 Ω at 77˚F
Temperature Coefficient	TEM-161-1 through -8	Positive, approximately 3 Ω/˚F
	TEM-161-54, -55	Positive, approximately 4.3 Ω/˚F
Tolerance Resistances	TEM-161-1, -2, -8	±1.0% at 70˚F
	TEM-161-3	±1.0% at 70˚F
	TEM-161-5	±1.0% at 70˚F
	TEM-161-54, -55	±0.05% at 77˚F
Set Point Range	TEM-161-8	55 to 85˚F
	TEM-161-54	55 to 85˚F

ACCESSORIES			
Code Number	**Description**		
WZ-100-4	Stainless steel immersion well for use with TEM-161-3	TEM-161-62	Toggle switch for use with TEM-161-54, -55, -4
TEM-161-61	Pushbutton switch for use with TEM-161-54, -55, -4	TEM-180-96	Electrical wall box mounting adapter kit includes wallplate adapter, mounting bracket, and screws

10. _____ If an 8′ averaging element sensor is required for an application, which sensor should be selected?

11. _____ Which sensor should be selected for a base room thermostat without a cover but with a setpoint?

12. _____ Which type of sensor is it?

13. _____ If a pushbutton accessory is needed for a room thermostat, what number should be selected?

14. _____ In the specification chart, what is the sensor type for a TEM 161-54 sensor?

15. _____ What is the reference resistance of the TEM 161-54 sensor?

16. _____ What is the temperature coefficient of the TEM 161-54 sensor?

17. _____ What is the resistor tolerance of the TEM 161-54 sensor?

Activity 17-2. Digital Inputs

Use the panel board wiring diagram to answer the questions.

1. _____ How many digital (binary) inputs are used in the system?

2. List three digital inputs.

_____

_____

3. _____ Are any electric meters indicated?

4. What is the purpose of filter-S?

_____

_____

5. _____ The part number for PS-1 is ___.

6. _____ The part number for PS-2 is ___.

7. _____ What type of reset do PS-1 and PS-2 have?

8. Why do they have this type of reset?

_____

_____

Use the hot water temperature control information to answer the questions.

Hot Water Temperature Control

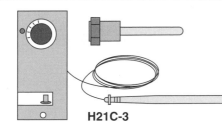

H21C-3

Description

This is a universal replacement for open high or SPDT applications. The control is furnished with a well assembly for 1/2″ tapping.

Applications

- operating control for hot water boilers

Features

- liquid-filled element provides rapid response to temperature change
- adjustable differential

SELECTION CHART							
Code Number	Application	Switch Action	Range*	Differential*	Well Connection Size—NPT	Range Adjuster	Maximum Bulb Temperature*
H21C-1	Open High (R-B) Open Low (R-Y)	SPDT	100 to 240	6 to 24	1/2 in.	Convertible	250
H21C-2		SPDT	100 to 240	6 to 24	1/2 in. 8 ft Cap.	Convertible	290
H21D-3	High Temp Lockout	Open High with Lockout	100 to 240	Manual Reset (locks out high)	1/2 in.	Knob	250

* in °F

9. _____ The code number of the control that requires a manual reset is ___.

10. _____ This control's maximum bulb temperature is ___°F.

11. _____ Is a separate well assembly required, or is it provided?

12. _____ The switch action of control H21C-2 is ___.

13. _____ On control H21C-2, terminals ___ open on a drop in temperature.

14. _____ Control H21C-2 range is ___°F.

15. _____ Control H21C-2 differential is ___°F.

Activity 17-3. Analog Outputs

Use the panel board wiring diagram to answer the questions.

1. _____ How many analog outputs are used in the system?

2. List three analog outputs.

_____

3. List the wire colors that are shown connected to the AHU controller.

_____

Use the rooftop air handling unit drawing to answer the questions.

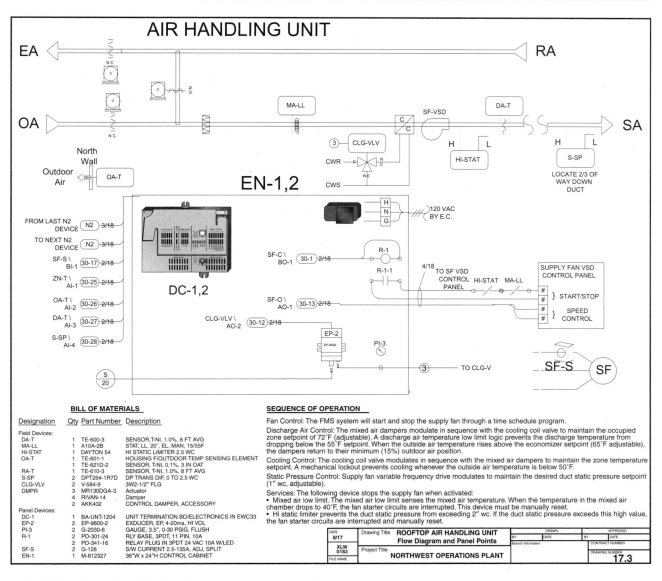

4. What does the SF-O do?

_____

_____

5. What is EP-2?

_____

_____

6. _____ What device is EP-2 connected to?

7. _____ Would the cooling valve have a spring range?

8. How do you know?

_____

_____

9. If only one analog output is used for three damper actuators, what does this indicate?

_____

_____

10. _____ Must the controller programming be consistent with the type of sensor used in the application?

Use the proportional valve actuator information to answer the questions.

Proportional Valve Actuator

Description

The PVA-122 proportional valve actuator is an electric actuator that provides proportional control valves with up to a 5/16″ stroke in HVAC applications.

Features

- simplified setup and adjust procedures decrease installation costs
- wide range of control inputs meets the needs of most applications
- Light Emitting Diode (LED) reduces commissioning time and displays operating status

SPECIFICATIONS		
Power Requirements	24 VAC at 50/60 Hz, 5.0 VA supply minimum, Class 2	
Input Signal	Jumper Selectable:	0 to 10 VDC, 0 to 20 VDC, 6 to 9 VDC, 0 to 20 mA
	Factory Setting:	0 to 10 VDC, Drive Down action on signal increase
Input Impedance	Voltage Input:	0 to 10 VDC, 150,000 Ω and 0 to 20 VDC, 450,000 Ω
	Current Input:	0 to 20 mA, 500 Ω
Force	Shutoff and Breakaway: 22 lb minimum	
Cycles	60,000 full cycles; 1,500,000 repositions	
Enclosure	NEMA 1	
Ambient Operating Conditions	35 to 135°F Maximum dew point at 90% RH, non-condensing	

11. _____ What is the factory setting for the input signal?

12. _____ What are the actuator power requirements?

13. What input signals may be selected with a jumper?

_____

_____

14. _____ What is the shutoff force?

15. _____ What displays the operating status of the controller?

Activity 17-4. Digital Outputs

Use the panel board wiring diagram to answer the questions.

1. _____ How many binary outputs are used in the system?

2. List the binary outputs.

   _____

3. A number of binary outputs are connected to an output relay. Why?

   _____

   _____

4. _____ Is any lighting control indicated?

5. _____ Are any incremental actuators indicated?

6. _____ How many binary outputs are used per incremental output?

Use the electric valve actuator information to answer the questions.

Electric Valve Actuator

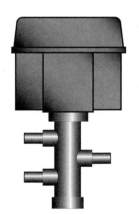

Description

The EVA-850 synchronous, motor-driven actuator provides floating control of valves with up to 3/4″ stroke in HVAC applications. This compact, non-spring return actuator has a 50 lb force and requires a 3-wire, 24 VAC signal from the controller.

SPECIFICATIONS	
Product	EVA-850-1 Electric Valve Actuator Assemblies
Control Mode	Floating Control, 3-Wire
Supply Voltage	24 VAC +6 V, −4 V, <200 mA, 50/60 Hz
Power Consumption	6 VA
Shutoff Force	50 lb Force Minimum
Stroke Time	1/2 in. Stroke: Approx. 65 Sec 3/4 in. Stroke: Approx. 90 Sec
Ambient Operating Temperature	0 to 140˚F, 10 to 90% RH Non-Condensing, 85˚F Max Dew Point
Media Temperature	Water: 190˚F Steam: 280˚F

Operation

A controller sends 24 VAC to the up or down terminal on the circuit board depending on the desired movement of the valve. This signal causes the motor to rotate in the proper direction, moving the valve up or down. When the controller stops sending a signal, the valve remains in place.

When the controller closes the valve, a shutoff force builds. When this force reaches 50 lb, the lever activates a force sensor which stops the motor.

7. _____ The maximum stroke for the actuator is ___″.

8. _____ Does the actuator have a spring return?

9. _____ What type of control does it support?

10. What happens to the valve when the controller stops sending a signal?

_____

_____

11. _____ The actuator power consumption is ___ VA.

12. _____ The approximate stroke time for a ½″ stroke valve is ___ sec.

Name: _____ Date: _____

Control System Principles

_____ **1.** All building automation system controllers should be wired in accordance with the ___ and local regulations.

    A. Occupational Safety and Health Administration (OSHA)
    B. American National Standards Institute (ANSI)
    C. National Electrical Code® (NEC®)
    D. International Electrotechnical Commission (IEC)

_____ **2.** A(n) ___ is a flat rail attached to a control panel and used for mounting.

_____ **3.** Operator interfaces commonly require installation ___″ above the finished floor for viewing by maintenance personnel.

    A. 48
    B. 60
    C. 72
    D. 84

T   F **4.** The most common building automation system controller power supply is a 230 VAC to 120 VAC step-down transformer.

_____ **5.** A(n) ___ is a circuit that has more than one point connected to earth ground, with a voltage potential difference between the two ground points high enough to produce a circulating current in the ground system.

_____ **6.** One large transformer is recommended for supplying multiple building automation system controllers when the controllers are located ___.

_____ **7.** ___ is a multipoint communication standard that incorporates low-impedance drivers and receivers providing high tolerance to noise.

    A. RS-485
    B. dip switch
    C. peripheral wiring
    D. interface cabling

_____ **8.** A(n) ___ is a configuration in which multiple controllers are connected in series.

T   F **9.** A network address is a unique number assigned to each building automation system controller on a communication network.

_____ **10.** Optoisolation is a communication method in which controllers use ___ components to prevent communication problems.

    A. series
    B. parallel
    C. photonic
    D. incremental

_____ **11.** Digital inputs are commonly configured as ___.

_____ **12.** Controller analog output terminals are commonly labeled ___, and COM.

    A. V1, V2, etc.
    B. T1, T2, etc.
    C. DO 1, DO 2, etc.
    D. AO 1, AO 2, etc.

_____ **13.** A(n) ___ is a device that is connected to a personal computer or building automation system controller to perform a specific function.

    A. incremental output device
    B. jumper
    C. normally open contact
    D. peripheral device

_____ **14.** A(n) ___ is a device that transmits multiple bits of information (0s and 1s) simultaneously.

T   F   **15.** A dumb terminal is a display monitor and keyboard, with limited processing capabilities.

_____ **16.** When testing communication wiring, such as RS-485+ to RS-485–, the voltage measurement should be between ___.

    A. 3 VDC and 5 VDC
    B. 15 VAC and 30 VAC
    C. 1 mV and 10 mV
    D. 4 mV and 20 mV

_____ **17.** The resistance of a negative temperature coefficient sensor ___ as the temperature increases.

_____ **18.** The ___ of a thermistor sensor is the amount of resistance change per degree Fahrenheit.

_____ **19.** A common thermistor nominal value is 1000 Ω at ___°F.

    A. 50
    B. 70
    C. 90
    D. 110

_____ **20.** A(n) ___ is a conductor used to connect pins on a controller or device.

Building Automation System Network Configuration

_____ **1.** multidrop

_____ **2.** star

_____ **3.** daisy chain

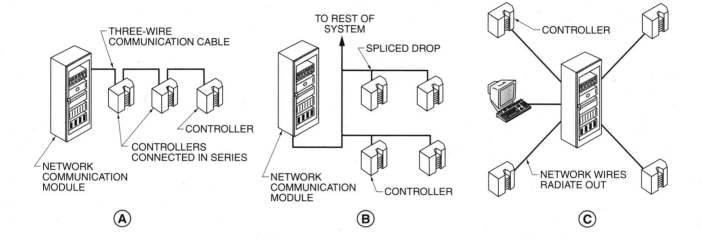

Controller Jumper Usage

_____ **1.** analog output voltage selection jumper

_____ **2.** analog output current selection jumper

_____ **3.** analog input current selection jumper

_____ **4.** jumper placement location

_____ **5.** analog input voltage/resistance selection jumper

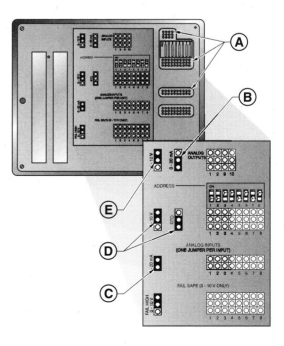

Network Communication Module Modem Connection

_____ **1.** serial cable

_____ **2.** modem

_____ **3.** network communication module power supply

_____ **4.** transformer

_____ **5.** building automation system local area network

_____ **6.** modem power supply

_____ **7.** unitary controller

_____ **8.** modem power cable

_____ **9.** phone line

_____ **10.** low-voltage (24 VAC) power line

_____ **11.** network communication module

_____ **12.** high-voltage (120 VAC) power lines

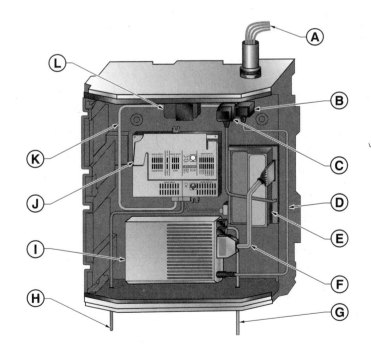

Building Automation System Installation, Wiring, and Testing

Name: _____   Date: _____

Activity 18-1. Communication Bus, Peripherals, and Power Wiring

Use the building automation system wiring legend to answer the questions.

BUILDING AUTOMATION SYSTEM WIRING LEGEND

CABLE/WIRE SPECIFICATION			TERMINALS PER CONDUCTOR COLOR			
Usage	Part Number	Description	Black	White	Jacket Color	Red
AI	CBL-18/2YEL	18/2 Shld Yellow	AI Com	———	AI	———
AI	CBL-18/3YEL	18/3 Shld Yellow	AI Com	Power	AI	———
AO	CBL-18/2TAN	18/2 Shld Tan	AO Com	———	AO	———
AO	CBL-18/3TAN	18/3 Shld Tan	AO Com	Power	AO	———
BI	CBL-18/2ORG	18/2 Shld Orange	BI 24V	———	BI	———
BO	CBL-18/2VLT	18/2 Shld Violet	BO Com	———	BO	———
BO	CBL-18/3VLT	18/3 Shld Violet	BO Com	BO	BO	———
GENERAL PURPOSE	CBL-18/2NAT	18/2 Shld Natural	Common	———	———	Power
GENERAL PURPOSE	CBL-18/3NAT	18/3 Shld Natural	Common	———	Signal	Power
CONTROLLER	CBL-24/8NAT	24/8 Natural	———	———	———	———
CONTROLLER PHONE JACK	CBL-STAT25	Pre-Term'd Blue	———	———	———	———
24 VAC	CBL-18/2GRY	18/2 Shld Grey	24 V Com	———	24 VAC	———
24 VAC POWER BUS	PB0137	14/2 Unsh White	Common	———	———	24 VAC
600 V	CBL-18/2600	18/2	———	———	———	———
600 V	CBL-18/3600	18/3	———	———	———	———
N2 BUS	CBL-18/3BLU	18/3 Shld Blue	N2-	N2-	N2+	———
N1 BUS-ARCNET	CBL-RG62PUR	RG62 Purple	———	———	———	———
N1 BUS-ETHERNET	63609	24/8 Purple	———	———	———	———
BACNET BUS	63609	24/8 Purple	———	———	———	———
LON BUS	43701	22/2 Shld Blue	———	———	———	———
NT BUS	00-4340	22/4 Shld Blue	———	———	———	———
XT BUS	CBL-18/3BLU	18/3 Shld Blue	XT-	XT-	XT+	———

ETHERNET PATCH CABLE

Pin No.	Color
1	WHT/ORN
2	ORN/WHT
3	WHT/WHT
4	BLU/WHT
5	WHT/BLU
6	GRN/WHT
7	WHT/BRN
8	BRN/WHT

ETHERNET CROSSOVER CABLE

Pin No.	Color
1	WHT/GRN
2	GRN/WHT
3	WHT/ORG
4	BLU/WHT
5	WHT/BLU
6	ORN/WHT
7	WHT/BRN
8	BRN/WHT

1. What is the N1 bus Ethernet description?

_____

_____

2. _____ What is the 24 VAC power bus part number?

3. _____ For an Ethernet patch cable, what pin number is associated with the WHT/GRN colors?

Activity 18-2. Input Identification

Use detail B20 and B22 to answer the questions.

1. List the wiring terminals and wire colors.

_____

_____

2. _____ If the resistance increases as the temperature increases, what kind of sensor is it?

3. _____ Are any jumpers involved for the setup of this AI?

4. _____ In detail B20, is the power supply polarity indicated?

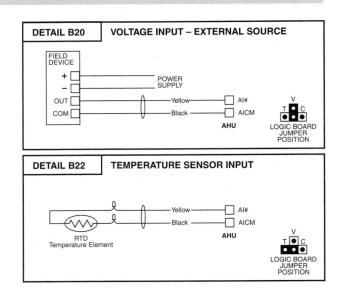

Activity 18-3. Binary Inputs

Use the panel board wiring diagram to answer the questions.

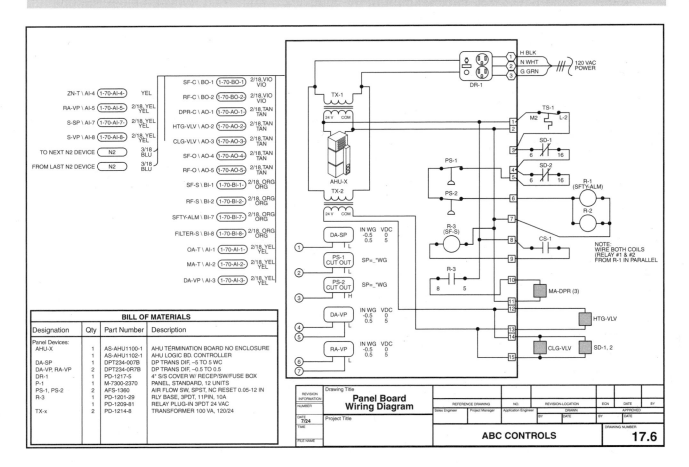

1. _____ To which binary input is the filter-S wired?

2. List the number of wires, wire gauge, and colors used for this binary input.

Use the building automation system wiring legend to answer the question.

3. List the part number(s) and description(s) of all binary input cable.

Activity 18-4. Binary Outputs

Use the panel board wiring diagram to answer the questions.

1. _____ To which binary output is RF-C wired?

2. List the number of wires, wire gauge, and colors used for this binary output.

Use the building automation system wiring legend to answer the question.

3. List the part number(s) and description(s) of all binary output cable.

Activity 18-5. Analog Outputs

Use the panel board wiring diagram to answer the question.

1. List the analog outputs, their use, and wire colors.

Use detail B24 to answer the questions.

2. _____ What wire color(s) is/are used at the analog outputs?

3. Where are they terminated?

_____

_____

4. _____ Is the polarity of the field device given?

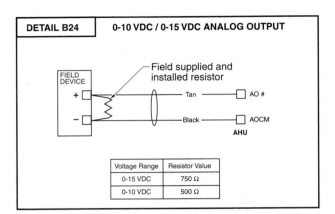

DETAIL B24	0-10 VDC / 0-15 VDC ANALOG OUTPUT

Voltage Range	Resistor Value
0-15 VDC	750 Ω
0-10 VDC	500 Ω

Use the building automation system wiring legend to answer the question.

5. List the part number(s) and description(s) of all analog output cable.

_____

_____

Activity 18-6. Temperature Sensor Resistance/Temperature Characteristics

Use the temperature sensor resistance/temperature characteristic charts to answer the questions.

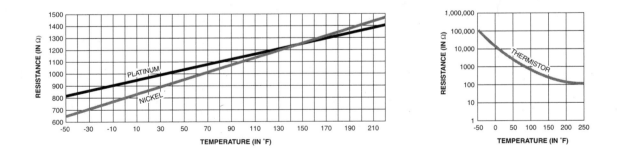

1. _____ The approximate resistance value for a platinum sensor at a temperature of 80°F is ___ Ω.

2. _____ The approximate resistance value for a nickel sensor at a temperature of 50°F is ___ Ω.

3. _____ The approximate resistance value for a thermistor sensor at a temperature of 150°F is ___ Ω.

Activity 18-7. Point Identification

Identify each point on the air handling unit as an analog input, analog output, binary input, or binary output.

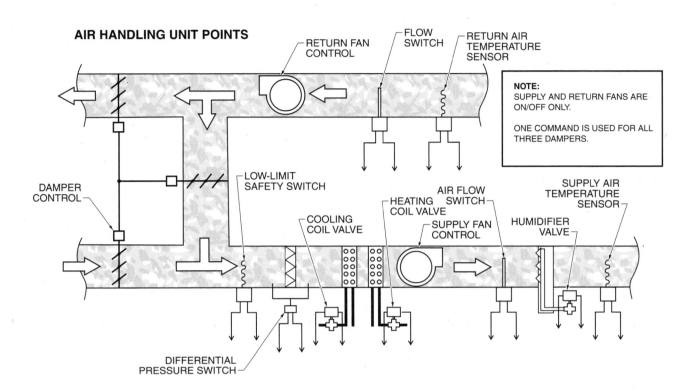

1. _____ Return fan control

2. _____ Flow switch

3. _____ Return air temperature sensor

4. _____ Supply air temperature sensor

5. _____ Humidifier valve

6. _____ Air flow switch

7. _____ Supply fan control

8. _____ Heating coil valve

9. _____ Cooling coil valve

10. _____ Differential pressure switch

11. _____ Low-limit safety switch

12. _____ Damper control

Name: _____ Date: _____

Control System Principles

T   F   **1.** A concentrator is interconnected equipment used for sending and receiving information.

_____   **2.** A ___ is a device, such as a computer or a printer, that has a unique address and is attached to a network.

      A. node
      B. topology
      C. router
      D. thinnet

_____   **3.** A(n) ___ normally encompasses one business or building and does not require the linking of a large number of users in remote locations.

_____   **4.** Network topology is the ___ of the network configuration.

_____   **5.** Ethernet is a local area network architecture that can connect up to ___ nodes and supports minimum data transfer rates of 10 megabits per second (Mbps).

      A. 256
      B. 512
      C. 1024
      D. 2048

_____   **6.** In an Ethernet network, data is divided into ___ before being transmitted.

      A. receipt notices
      B. concentrators
      C. bundles
      D. packets

_____   **7.** A(n) ___ occurs when two nodes transmit simultaneously and data is corrupted.

T   F   **8.** An adapter card (network interface card) is a card (circuit board) installed in a network component to allow it to communicate with the network.

_____ **9.** A(n) ___ is a network switchboard that allows a number of nodes to communicate with each other.

_____ **10.** A switch manages the communication between networks or parts of a network that operate at ___ data transmission speeds.

    A. low
    B. high
    C. different
    D. the same

_____ **11.** A(n) ___ manages the communication between different networks.

_____ **12.** ___ is the cables and other network devices, such as hubs and switches, that make up a network.

    A. Hub
    B. Thicknet
    C. Physical layer
    D. Ping

T   F   **13.** Submasking is a data transmission method where data is sent without frequency modulation.

_____ **14.** A(n) ___ is a cable used for connections from node to node.

_____ **15.** ___ is thick coaxial cable classified as 10base5.

_____ **16.** The most common personal computer operating system used today is ___.

    A. Microsoft® Windows®
    B. Ethernet
    C. Macintosh℠
    D. OS10

_____ **17.** An Internet Protocol address consists of ___.

_____ **18.** A ___ is an address that can change at any time.

    A. static address
    B. dynamic address
    C. mixed network address
    D. all of the above

T   F   **19.** Submasking is an addressing scheme that filters messages and determines if the message is to be passed to local nodes on a subnetwork or if it is to be sent on to the main network through the router.

_____ **20.** A(n) ___ is an echo message and its reply sent by one network device to detect the presence of another device.

T   F   **21.** Each device on the Internet has an address known as the Media Access Control (MAC) address.

_____ **22.** A MAC address is ___ into the device by the manufacturer.

_____ **23.** The ___ command displays the computer name on the network.

     A. ipconfig
     B. ping
     C. hostname
     D. none of the above

T   F   **24.** Winipcfg is a graphical option of the ipconfig command.

_____ **25.** A(n) ___ acts as a central device that connects the computers to each other and to the network or the Internet.

_____ **26.** A(n) ___ network is a network in which new Ethernet network components are added to early Arcnet network components.

     A. intranet
     B. mixed
     C. Bluetooth
     D. all of the above

T   F   **27.** Bluetooth technology is used to allow short-range communication between devices for checkout and commissioning.

_____ **28.** The Bluetooth device operating range depends on the device class and can range from 1′ up to ___′.

     A. 10
     B. 50
     C. 150
     D. 300

T   F   **29.** Virtual private network software may be installed to allow remote access to a facility network using encryption and security techniques.

_____ **30.** Security for web-enabled control systems is provided by the network ___, which does not allow access from the outside.

Ethernet Network Star Configurations

_____ **1.** local communication bus

_____ **2.** operator workstation

_____ **3.** controller

_____ **4.** central device (network communication module)

_____ **5.** Ethernet local area network

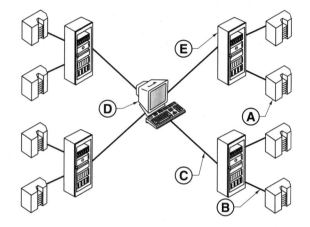

Ethernet Network Common-Bus Configurations

_____ **1.** controller

_____ **2.** central device (network communication module)

_____ **3.** Ethernet local area network

_____ **4.** operator workstation

_____ **5.** local communication bus

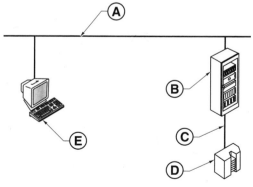

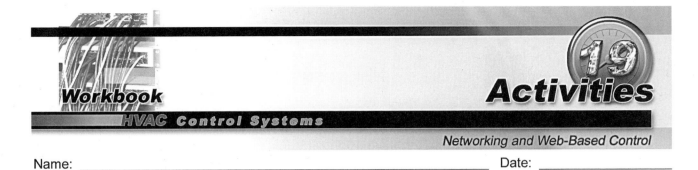

Activity 19-1. Building Automation System Troubleshooting

A building automation system does not appear to be communicating properly on the network. Before the manufacturer is called for service, basic tests are run on the system. Use the communication riser drawing to answer the questions.

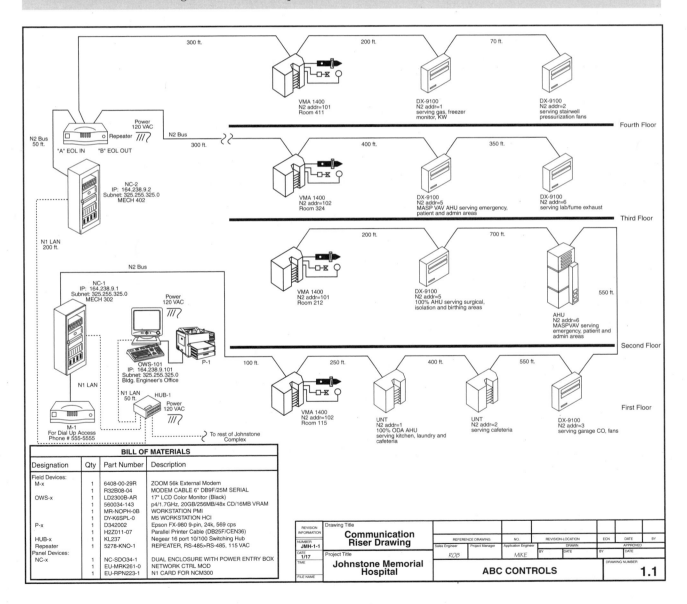

1. _____ Complete the command line to determine if the OWS-101 NIC card is operating properly. C:\>ping ___

2. _____ Complete the command line to determine if NC-1 is communicating properly. C:\>ping ___

3. _____ Complete the command line to determine if NC-2 is communicating properly. C:\>ping ___

4. _____ The subnet mask for OWS-101 is ___.

5. _____ The subnet mask for NC-1 is ___.

6. _____ The subnet mask for NC-2 is ___.

7. _____ Are the subnet masks for OWS-101, NC-1, and NC-2 the same?

8. _____ Consideration has been given to expanding the building automation system by adding another workstation. A possible IP address for the new workstation OWS-102 is ___.

9. _____ A possible subnet address for the new workstation OWS-102 is ___.

10. _____ Consideration has been given to expanding the building automation system by adding another NC. A possible IP address for the new NC, NC-3 is ___.

11. _____ A possible subnet address for the new NC, NC-3 is ___.

Name: _____ Date: _____

Control System Principles

_____ **1.** A(n) ___ is a building automation system software method used to control the energy-using equipment in a building.

T   F   **2.** A direct digital control system is a control system in which the building automation system controller is wired directly to controlled devices and can turn them ON and OFF, or start a motor.

_____ **3.** ___ control is control in which feedback occurs between the controller, sensor, and controlled device.

        A. Open loop
        B. Closed loop
        C. High/low signal
        D. Universal input-output

_____ **4.** Feedback is the measurement of the results of a controller action by a(n) ___.

_____ **5.** The most common direct digital control feature is the ability to ___ in a building automation system.

_____ **6.** ___ is the difference between a control point and a setpoint.

        A. Offset
        B. Lead/lag
        C. Algorithm
        D. Tuning

_____ **7.** Setback is the unoccupied ___ setpoint.

        A. cooling
        B. heating
        C. humidity
        D. static pressure

_____ **8.** ___ is a direct digital control feature in which a primary setpoint is reset automatically as another value (reset variable) changes.

T   F   **9.** When using the dry bulb economizer method, an outside air temperature sensor is used to control economizer operation.

_____ **10.** ___ is the alternation of operation between two or more similar pieces of equipment.

_____ **11.** ___ is a direct digital control feature in which a building automation system selects among the highest or lowest values from multiple inputs.

_____ **12.** ___ control is a direct digital control feature that calculates an average value from all selected inputs.

    A. Averaging
    B. Proportional
    C. Integral
    D. Derivative

_____ **13.** An algorithm is a mathematical equation used by a building automation system controller to determine a desired ___.

_____ **14.** A(n) ___ is a control algorithm that positions the controlled device in direct response to the amount of offset in a building automation system.

_____ **15.** ___ control algorithms are used in basic control systems that do not require precise control.

    A. Averaging
    B. Proportional
    C. Integral
    D. Derivative

T   F   **16.** An integral control algorithm is a control algorithm that eliminates any offset after a certain length of time.

_____ **17.** In the HVAC industry, only extremely sensitive control applications require ___ control.

_____ **18.** A(n) ___ is a control algorithm that automatically adjusts its response time based on environmental conditions.

_____ **19.** ___ is the downloading of the proper response times into a controller and checking the response of the control system.

    A. Overshooting
    B. Undershooting
    C. Strategies
    D. Tuning

_____ **20.** When using a rooftop packaged unit direct digital control system, a(n) ___ is used in place of a standard thermostat control.

Proportional Control

_____ **1.** setpoint

_____ **2.** offset

_____ **3.** control point

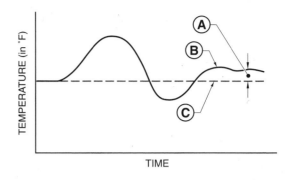

Unitary Control of Rooftop Packaged Unit

_____ **1.** building space temperature sensor

_____ **2.** supply fan

_____ **3.** return air

_____ **4.** air flow switch

_____ **5.** compressor

_____ **6.** unitary controller

_____ **7.** heating element

_____ **8.** supply air flow

_____ **9.** filter

_____ **10.** air flow switch paddle

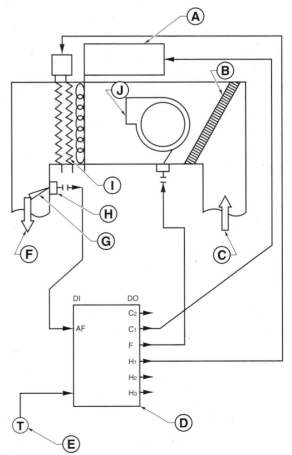

Air Handling Unit Control of Variable Air Volume System

_____ **1.** normally open damper

_____ **2.** static pressure sensor

_____ **3.** return fan

_____ **4.** air flow station with electronic transducer

_____ **5.** variable-speed drive

_____ **6.** supply air

_____ **7.** supply fan

_____ **8.** air handling unit controller

_____ **9.** normally closed damper

_____ **10.** outside air

_____ **11.** return air

_____ **12.** exhaust air

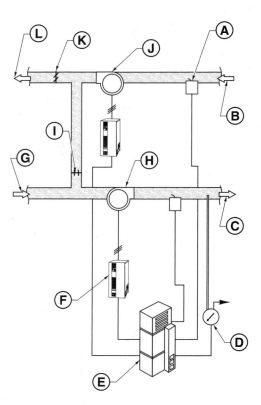

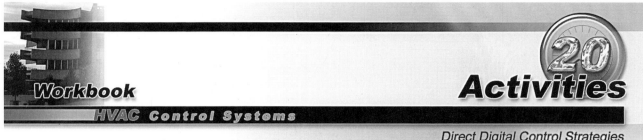

Workbook

HVAC Control Systems

Direct Digital Control Strategies

Name: _____ Date: _____

Activity 20-1. Rooftop Air Handling Unit Control Strategies

Use the rooftop air handling unit sequence of operation to answer the questions.

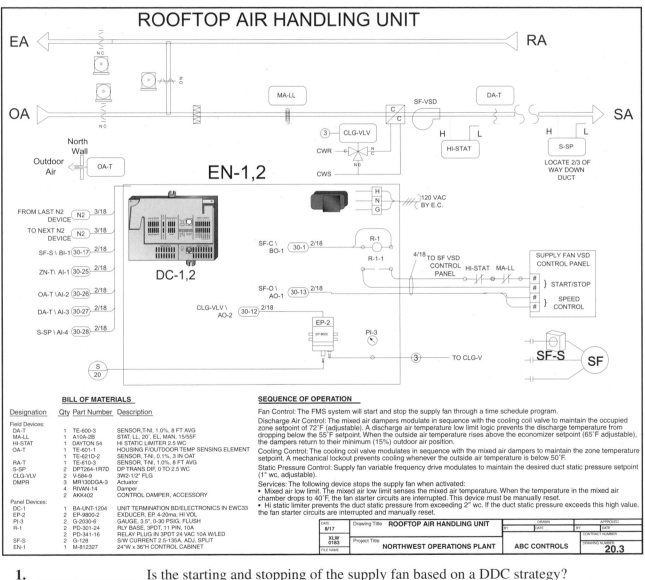

1. _____ Is the starting and stopping of the supply fan based on a DDC strategy?

2. _____ If it is not based on temperature, what is it based on?

3. _____ The occupied zone setpoint is ___°F.

4. _____ Is the occupied zone setpoint adjustable?

5. What devices are modulated in sequence to maintain the cooling setpoint?

_____

6. _____ What DDC strategy prevents the discharge air temperature from dropping too low?

7. _____ The discharge air temperature setpoint is ___°F.

8. _____ What DDC strategy determines if outside air can provide cooling?

9. _____ What type of DDC strategy is it?

10. _____ The outside air setpoint is ___°F.

11. How can it be determined that humidity control is not involved?

_____

_____

12. What DDC strategy maintains static pressure in the supply duct?

_____

13. _____ The static pressure setpoint is ___.

14. _____ Is the static pressure setpoint adjustable?

15. _____ Is the mixed air low limit connected to the rooftop controller?

16. _____ Is any averaging or reset control indicated?

Activity 20-2. Rooftop Air Handling Unit Service Call

A too-hot complaint is received from the area supplied by the rooftop air handling unit. In addition, the occupants claim that the unit is often noisy. After investigation, it is discovered that the duct static pressure is hunting (cycling) and not maintaining setpoint. Duct noise increases as the static pressure increases. A laptop computer or service tool is connected to the controller.

1. List the components of the closed loop control for the fan static pressure.

_____

After examination, it is determined that the static loop parameters were downloaded but never checked.

2. _____ What might the static pressure control loop need?

3. List the basic steps of how this might be done.

_____

_____

Activity 20-3. Hot Water Boiler Control Strategies

Use the hot water boiler control sequence of operation to answer the questions.

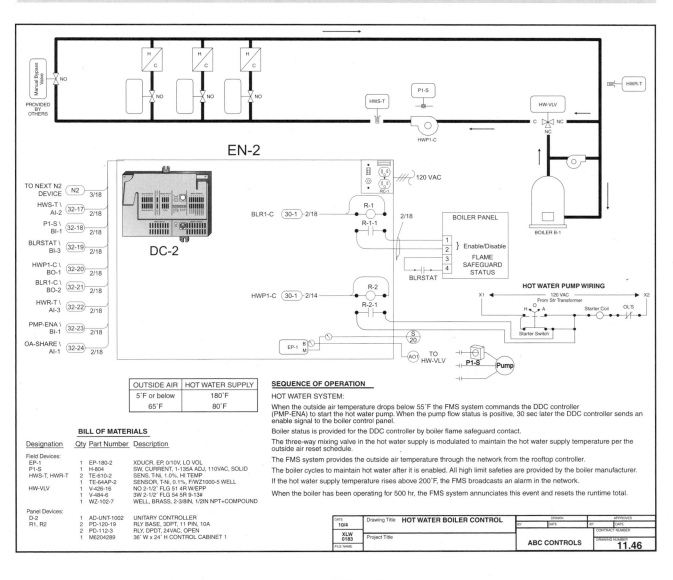

OUTSIDE AIR	HOT WATER SUPPLY
5°F or below	180°F
65°F	80°F

SEQUENCE OF OPERATION

HOT WATER SYSTEM:

When the outside air temperature drops below 55°F the FMS system commands the DDC controller (PMP-ENA) to start the hot water pump. When the pump flow status is positive, 30 sec later the DDC controller sends an enable signal to the boiler control panel.

Boiler status is provided for the DDC controller by boiler flame safeguard contact.

The three-way mixing valve in the hot water supply is modulated to maintain the hot water supply temperature per the outside air reset schedule.

The FMS system provides the outside air temperature through the network from the rooftop controller.

The boiler cycles to maintain hot water after it is enabled. All high limit safeties are provided by the boiler manufacturer.

If the hot water supply temperature rises above 200°F, the FMS broadcasts an alarm in the network.

When the boiler has been operating for 500 hr, the FMS system annunciates this event and resets the runtime total.

BILL OF MATERIALS

Designation	Qty	Part Number	Description
Field Devices:			
EP-1	1	EP-180-2	XDUCR, EP, 0/10V, LO VOL
P1-S	1	H-804	SW, CURRENT, 1-135A ADJ, 110VAC, SOLID
HWS-T, HWR-T	2	TE-610-2	SENS, T-Ni, 1.0%, HI TEMP
	1	TE-64AP-2	SENSOR, T-Ni, 0.1%, F/WZ1000-5 WELL
HW-VLV	1	V-426-16	NO 2-1/2" FLG 51 4R W/EPP
	1	V-484-6	3W 2-1/2" FLG 54 5R 9-13#
	1	WZ-102-7	WELL, BRASS, 2-3/8IN, 1/2IN NPT+COMPOUND
Panel Devices:			
D-2	1	AD-UNT-1002	UNITARY CONTROLLER
R1, R2	2	PD-120-19	RLY BASE, 3DPT, 11 PIN, 10A
	2	PD-112-3	RLY, DPDT, 24VAC, OPEN
	1	M6204289	36" W x 24" H CONTROL CABINET 1

DATE 10/4	Drawing Title **HOT WATER BOILER CONTROL**		DRAWN		APPROVED	
			BY	DATE	BY	DATE
XLW 0183	Project Title				CONTRACT NUMBER	
FILE NAME			**ABC CONTROLS**		DRAWING NUMBER **11.46**	

1. _____ What starts and stops pump HWP1-C?

2. _____ What is its setpoint?

3. _____ What device provides boiler status?

4. _____ What device is modulated to maintain the correct outside air temperature in the system?

5. _____ Does the hot water supply temperature have a fixed setpoint?

6. _____ What DDC strategy changes the hot water setpoint as the outside air temperature changes?

7. _____ The reset schedule outside air temperature values are ___°F and ___°F.

8. _____ The reset schedule hot water supply temperature values are ___°F and ___°F.

9. _____ Where does the outside air temperature value come from?

10. _____ Is there an alarm setpoint?

11. _____ If so, the alarm setpoint value is ___°F.

Activity 20-4. Hot Water Boiler Service Call

A too-cold complaint is received from the area supplied by the hot water boiler. After investigation, it is determined that the area is too cold. The hot water pump and boiler are running, and the hot water supply temperature is warm. A laptop computer or service tool is connected to the controller.

1. List the components of the closed loop control for the hot water supply temperature.

_____

_____

The outside air temperature is checked and determined to be 35°F. The hot water supply setpoint is given as 150°F.

2. _____ Is the hot water supply setpoint correct?

3. _____ Using the reset schedule on the hot water boiler control print, an approximate hot water setpoint at the given outside air temperature is ___°F.

Upon checking the reset schedule in the software, it is discovered that the wrong values were entered. After correcting the values, it is noticed that the hot water three-way valve is constantly opening and closing.

4. List three things that may cause this condition.

_____

_____

Upon further checking, it is discovered that the integration value is 0.

5. Should this value be changed? Why or why not?

_____

_____

Name: _____ Date: _____

Control System Principles

_____ **1.** A(n) ___ is a programmable software method used to control the energy-consuming functions of a commercial building.

_____ **2.** ___ is a control strategy for life safety issues such as fire prevention, detection, and suppression.

      A. Time-based supervisory control
      B. Optimum start/stop supervisory control
      C. Duty cycling supervisory control
      D. Life safety supervisory control

T   F **3.** Optimum start supervisory control is a control strategy in which the time of day is used to determine the desired operation of a load.

_____ **4.** ___ allows an HVAC technician to individually program building automation system ON and OFF time functions for each day of the week.

_____ **5.** A(n) ___ is a time-based supervisory control strategy in which the occupants can change a zone from an unoccupied to occupied mode for temporary occupancy.

_____ **6.** ___ compensates for a specific event in a building without using a timed override.

      A. Schedule linking
      B. Saving time changeover
      C. Alternate scheduling
      D. Temporary scheduling

_____ **7.** Schedule linking provides the ability for building automation systems to join ___ that are used during the same time.

      A. controllers
      B. loads
      C. electrical demands
      D. floors

_____ **8.** ___ is a supervisory control strategy in which the HVAC load is turned ON as late as possible to obtain the proper building space temperature at the beginning of building occupancy.

_____ **9.** ___ is a control method that adjusts (learns) its control settings based on the condition of a building.

T F **10.** A shed ratio is the ratio of an indoor temperature change and the length of time it takes to obtain that temperature change.

_____ **11.** The optimum stop supervisory control strategy is commonly limited to a specific length of time, such as ___ min or ___ min.

   A. 5; 10
   B. 10; 20
   C. 15; 30
   D. 30; 60

_____ **12.** The duty cycling supervisory control strategy is designed to reduce ___ in a commercial building.

_____ **13.** ___ is the highest amount of electricity used during a specific period of time.

_____ **14.** A ___ is an electric load that has been turned OFF by an electrical demand supervisory control strategy.

   A. low-priority load
   B. targeted load
   C. restored load
   D. shed load

_____ **15.** A(n) ___ is a table that prioritizes the order in which electrical loads are turned OFF.

_____ **16.** A high-priority load is a load that is important to the operation of a building and is turned OFF ___ when demand goes up.

T F **17.** When building electrical demand is above the target, the loads in the low-priority shed table are shed in order.

_____ **18.** ___ is an electrical demand supervisory control strategy in which the order of loads to be shed is changed with each high electrical demand condition.

_____ **19.** A(n) ___ is a holiday that changes its date each year.

_____ **20.** ___ values can also be used as indicators of HVAC equipment efficiency and/or mechanical problems.

   A. Estimation control
   B. Duty cycling
   C. Life safety supervisory
   D. Thermal recovery coefficient

Duty Cycling

_____ **1.** load 2 ON

_____ **2.** one load ON at any given time

_____ **3.** load 1 ON

_____ **4.** load 2 OFF

_____ **5.** load 1 OFF

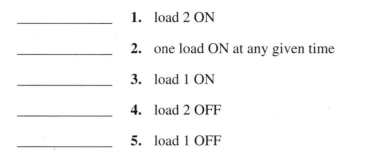

MULTIPLE LOADS

Maximum Shed and Minimum Shed

_____ **1.** shortest length of time load will operate

_____ **2.** minimum length of time load is OFF

_____ **3.** maximum length of time load is OFF

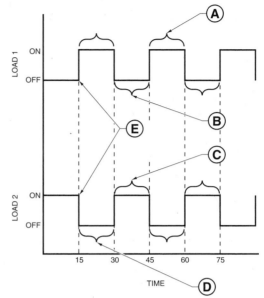

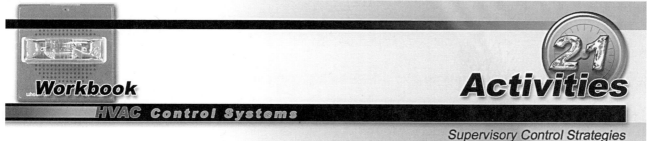

Name: _____   Date: _____

Activity 21-1. Rooftop Air Handling Unit Control Strategies

Use the rooftop air handling unit sequence of operation to answer the questions.

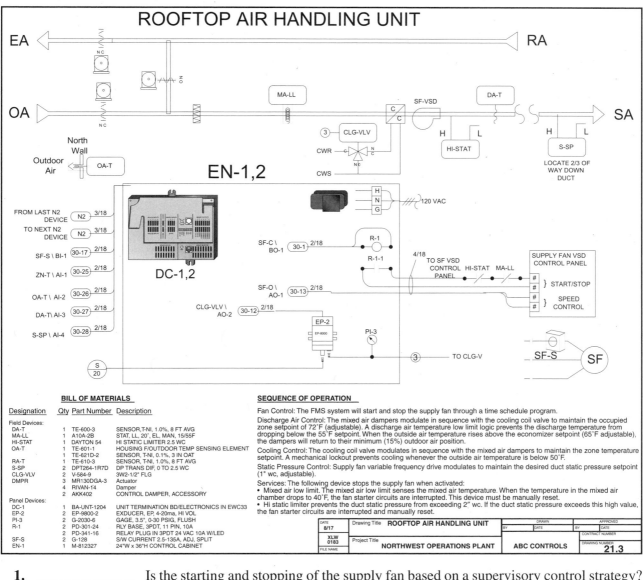

BILL OF MATERIALS

Designation	Qty	Part Number	Description
Field Devices:			
DA-T	1	TE-600-3	SENSOR,T-NI, 1.0%, 8 FT AVG
MA-LL	1	A10A-2B	STAT, LL, 20", EL, MAN, 15/55F
HI-STAT	1	DAYTON 54	HI STATIC LIMITER 2.5 WC
OA-T	1	TE-601-1	HOUSING F/OUTDOOR TEMP SENSING ELEMENT
	1	TE-621D-2	SENSOR, T-NI, 0.1%, 3 IN OAT
RA-T	1	TE-610-3	SENSOR, T-NI, 1.0%, 8 FT AVG
S-SP	2	DPT264-1R7D	DP TRANS DIF, 0 TO 2.5 WC
CLG-VLV	1	V-584-9	3W2-1/2" FLG
DMPR	3	MR130DGA-3	Actuator
	4	RIVAN-14	Damper
	2	AKK402	CONTROL DAMPER, ACCESSORY
Panel Devices:			
DC-1	1	BA-UNT-1204	UNIT TERMINATION BD/ELECTRONICS IN EWC33
EP-2	2	EP-9800-2	EXDUCER, EP, 4-20ma, HI VOL
PI-3	2	G-2030-6	GAGE, 3.5", 0-30 PSIG, FLUSH
R-1	2	PD-301-24	RLY BASE, 3PDT, 11 PIN, 10A
	2	PD-341-16	RELAY PLUG IN 3PDT 24 VAC 10A W/LED
SF-S	2	G-128	S/W CURRENT 2.5-135A, ADJ, SPLIT
EN-1	1	M-812327	24"W x 36"H CONTROL CABINET

SEQUENCE OF OPERATION

Fan Control: The FMS system will start and stop the supply fan through a time schedule program.

Discharge Air Control: The mixed air dampers modulate in sequence with the cooling coil valve to maintain the occupied zone setpoint of 72°F (adjustable). A discharge air temperature low limit logic prevents the discharge temperature from dropping below the 55°F setpoint. When the outside air temperature rises above the economizer setpoint (65°F adjustable), the dampers will return to their minimum (15%) outdoor air position.

Cooling Control: The cooling coil valve modulates in sequence with the mixed air dampers to maintain the zone temperature setpoint. A mechanical lockout prevents cooling whenever the outside air temperature is below 50°F.

Static Pressure Control: Supply fan variable frequency drive modulates to maintain the desired duct static pressure setpoint (1" wc, adjustable).

Services: The following device stops the supply fan when activated:
• Mixed air low limit. The mixed air low limit senses the mixed air temperature. When the temperature in the mixed air chamber drops to 40°F, the fan starter circuits are interrupted. This device must be manually reset.
• Hi static limiter prevents the duct static pressure from exceeding 2" wc. If the duct static pressure exceeds this high value, the fan starter circuits are interrupted and manually reset.

DATE 8/17	Drawing Title **ROOFTOP AIR HANDLING UNIT**		DRAWN		APPROVED	
			BY	DATE	BY	DATE
XLW 0183	Project Title				CONTRACT NUMBER	
FILE NAME	**NORTHWEST OPERATIONS PLANT**	**ABC CONTROLS**			DRAWING NUMBER **21.3**	

1. _____ Is the starting and stopping of the supply fan based on a supervisory control strategy?

2. _____ What supervisory control strategy is the starting and stopping of the supply fan based on?

3. _____ If the air handling unit were in a hospital, would this control strategy be used?

4. _____ What device located at the air handling unit uses electrical power?

5. What factors may be taken into consideration when considering demand control for the unit?

_____

_____

6. If the air handling unit static pressure setpoint were reduced, what would happen to the electrical demand at the supply fan?

_____

Activity 21-2. Rooftop Air Handling Unit Service Call

On Wednesday at 11 AM, a too-hot complaint is received from the area supplied by the rooftop air handling unit. Upon arrival, it is determined that the unit is off. A laptop computer or service tool is connected to the controller.

1. What supervisory control strategies may be keeping the unit off?

_____

_____

_____

After checking the software, it is determined that the time schedule was incorrectly set to turn the unit off at 10 AM instead of 10 PM. After changing the value, the unit starts.

2. What other time scheduling features may be checked to ensure that they are not incorrectly set?

_____

_____

_____

Activity 21-3. Single Duct VAV Box Control Strategies

Use the VAV box sequence of operation to answer the questions.

VAV BOX SEQUENCE OF OPERATION

UNOCCUPIED MODE:

THE VAV BOX IS PUT INTO UNOCCUPIED MODE BY A TIME SCHEDULE IN THE BUILDING DDC SYSTEM. IN THIS MODE, THE VAV BOX MAINTAINS THE BUILDING SPACE CONDITIONS TO THE SETUP/SETBACK TEMPERATURE SETPOINTS. THE BUILDING SPACE TEMPERATURE SENSOR MODULATES THE BOX DAMPER DEPENDING ON BUILDING SPACE REQUIREMENTS.

OCCUPIED MODE:

THE VAV BOX IS PUT INTO OCCUPIED MODE BY A TIME SCHEDULE IN THE BUILDING DDC SYSTEM. IN THIS MODE, THE VAV BOX MODULATES THE VARIABLE VOLUME DAMPER TO MAINTAIN THE BUILDING SPACE CONDITIONS AT THE OCCUPIED TEMPERATURE SETPOINTS.

ON A CALL FOR COOLING, THE VARIABLE VOLUME DAMPER IS MODULATED FROM MINIMUM AIR FLOW TO MAXIMUM COOLING AIR FLOW TO MAINTAIN BUILDING SPACE CONDITIONS. ON A DROP IN BUILDING SPACE TEMPERATURE, THE VARIABLE VOLUME DAMPER IS MODULATED TO ITS MINIMUM FLOW POSITION. ON A FURTHER CALL FOR HEAT, THE ELECTRIC HEAT CYCLES ON TO MAINTAIN SETPOINT.

SERIES FAN:

THE SERIES FAN IS OFF DURING THE SHUTDOWN MODE. THE FAN IS ALWAYS ON DURING THE OCCUPIED MODE AND IS CYCLED ON DURING THE UNOCCUPIED MODE.

THE ON/OFF SERIES FAN IS CONTROLLED BY A SINGLE BINARY OUTPUT WITH MINIMUM ON/OFF TIMERS THAT ARE ADJUSTABLE.

RADIANT HEATING CONTROL:

THE PERIMETER AREAS HAVE FINNED-TUBE RADIANT HEATING. THE CONTROL OF THE FINNED-TUBE HEATERS COMES FROM THE PERIMETER VAV CONTROLLERS. TYPICALLY, ON A CALL FOR HEAT THE DAMPER OF THE VAV CLOSES AND RECIRCULATES PRECONDITIONED BUILDING SPACE AIR. ON A CONTINUED CALL FOR HEAT, THE TWO-POSITION VALVE ON THE RADIANT FINNED TUBES OPENS AS THE BUILDING SPACE TEMPERATURE BECOMES SATISFIED, THE VALVES CLOSE.

1. _____ What supervisory strategy puts the VAV box into unoccupied mode?

2. What happens to the temperature setpoints?

_____

_____

3. What does the VAV box fan do during the occupied mode?

_____

4. What does the VAV box fan do differently during the unoccupied mode?

_____

_____

5. _____ If the VAV box controller has a minimum air flow setpoint of 200 cfm during the occupied mode, would the same amount of air be needed in the unoccupied mode?

6. _____ Does the system include electric heating elements?

Activity 21-4. Single Duct VAV Box Service Call

A too-hot complaint is received from the area supplied by a single duct VAV box. After investigation, it is determined that the room temperature is 85°F. A laptop computer or service tool is connected to the controller. The laptop or service tool indicates that the occupied cooling setpoint is 75°F, the unoccupied cooling setpoint is 85°F, and the current setpoint is 85°F.

1. _____ What mode is the VAV box in?

2. _____ What should be changed?

3. _____ Where should it be changed?

Name: _____   Date: _____

Control System Principles

_____ **1.** A(n) ___ is the process of upgrading a building automation system by replacing obsolete or worn parts, components, and controls with new or modern ones.

        A. commissioning
        B. retrofit
        C. download
        D. interface setup

_____ **2.** A building survey is an inventory of the ___ in a commercial building.

_____ **3.** ___ indicates the sequences of operation, lists of control points and locations, setpoints, strategies, and interlocks for the building automation system.

T   F   **4.** Programming is the creation of software.

_____ **5.** A(n) ___ is the completed programming of a controller.

_____ **6.** ___ is the checkout procedure used to start up a new building automation system.

        A. Commissioning
        B. Point mapping
        C. Interface setup
        D. System links

_____ **7.** A(n) ___ is the central computer that enables the building staff to view the operation of the building automation system.

_____ **8.** Point mapping is the process of adding the individual ___ to the database of the human-computer interface.

_____ **9.** Common alternate retrofit strategies for variable air handling units include ___.

        A. using a variable-frequency drive
        B. resetting the fan static setpoint based on zone demand
        C. removing fan inlet vane volume controls
        D. all of the above

_____ **10.** A(n) ___ is a controller that modulates the damper inside a VAV terminal box to maintain a specific building space temperature.

_____ **11.** ___ and ___ all hazards before beginning any retrofit.

T   F   **12.** A flame safeguard system is burner control equipment that monitors a burner start-up sequence and the main flame during normal operation, and provides an air purge to rid the combustion chamber of unburned fuel during a shutdown.

_____ **13.** Many hot water boiler systems have a ___ which controls the temperature of the water pumped through the building heating coils.

     A. pressure valve
     B. three-way valve
     C. 120 V control circuit
     D. digital interface

_____ **14.** ___ are used for many retrofit tasks, so efficient project management, planning, and coordination are required for a successful installation.

T   F   **15.** Components containing hazardous materials such as mercury must be disposed of properly, following all environmental regulations.

_____ **16.** A(n) ___ combines heating and cooling operations in one piece of equipment for most commercial buildings.

_____ **17.** A(n) ___ is an air handling unit that moves a variable volume of air.

T   F   **18.** Commissioning uses software to override inputs, outputs, and setpoints to ensure proper operation.

_____ **19.** Most new chillers have a built-in ___.

     A. cooling tower
     B. water circulation system
     C. microprocessor control panel
     D. power supply system

_____ **20.** When retrofitting a new building automation system to an older existing system, it is common to use a(n) ___ that allows the new building automation system to view and adjust the old building system.

Database Downloading

_____ **1.** local area network

_____ **2.** controller

_____ **3.** operator workstation

_____ **4.** communication bus

_____ **5.** network controller

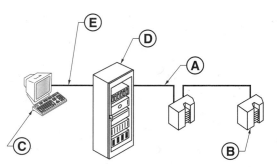

VAV Terminal Box

_____ **1.** terminal box

_____ **2.** actuator

_____ **3.** VAV terminal box controller

Terminal Box Retrofit

_____ **1.** differential pressure transducer

_____ **2.** controller

_____ **3.** VAV terminal box

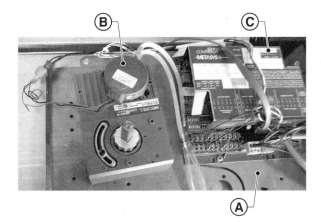

Review Questions 22

Name: _____ Date: _____

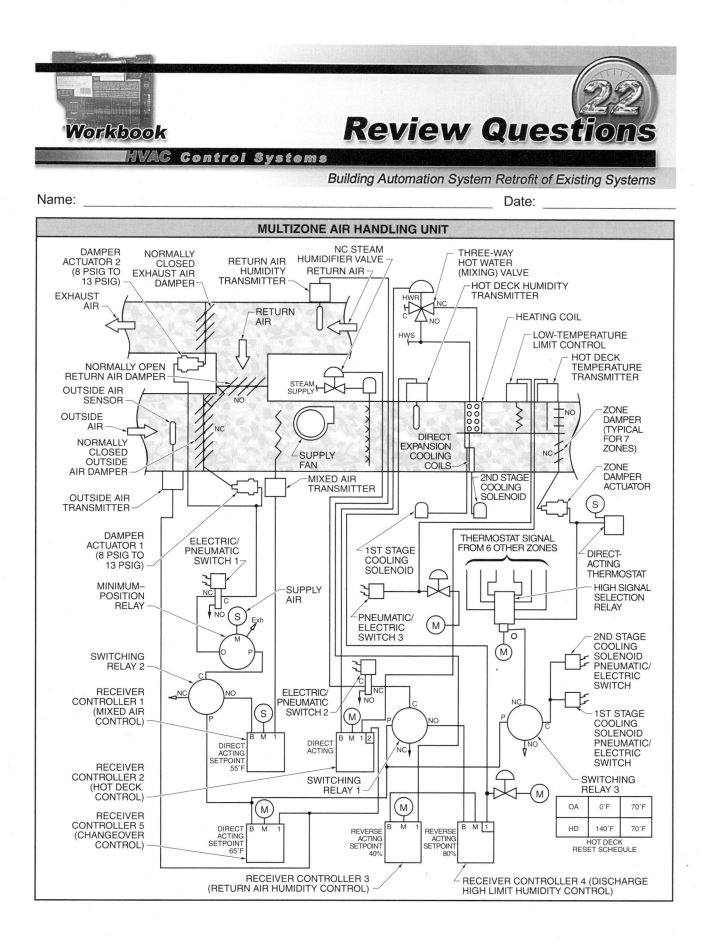

MULTIZONE AIR HANDLING UNIT

Activity 22-1. Mechanical System Identification

A building contains a number of pneumatically-controlled air handling units that must be retrofitted to direct digital control. Use the multizone air handling unit drawing to answer the questions.

1. _____ The air handling unit type is ___.

2. _____ The air handling unit supplies conditioned air to ___ (number) zone(s).

3. _____ The air handling unit contains ___ (number) fan(s).

4. List the air handling unit pneumatic actuators.

_____

_____

5. _____ The total number of actuators in the system is ___.

_____

_____

6. _____ The number of pneumatic temperature sensors is ___.

7. List the binary (on/off) devices in the system.

_____

8. _____ The control strategy needed for the hot deck is ___.

9. Two control strategies that may be used with the outside air dampers are ___.

_____

_____

10. _____ Two control strategies that may be used with the humidifier are ___.

11. The control strategy used to control the mechanical cooling is ___.

12. The number of electric/pneumatic transducers needed is ___.

13. _____ At a cost of $55 per transducer, the total transducer cost is $___.

14. _____ If 1 hr of installation time is needed per transducer, at $75/hr, the installation cost of the electric/pneumatic transducers is $___.

15. _____ If each temperature sensor costs $80, the total temperature sensor cost is $___.

Name: _____ Date: _____

Control System Principles

_____ 1. A(n) ___ provides information to maintenance technicians who maintain or manage a commercial building.

_____ 2. ___ is the most common building system management function used in commercial buildings.

_____ 3. The most common alarms used in commercial buildings are associated with ___ sensors.

     A. temperature sensors
     B. humidity sensors
     C. static pressure sensors
     D. flow sensors

T   F   4. Noncritical alarms concern devices vital to proper operation of a commercial building.

_____ 5. A(n) ___ is the amount of change required in a variable for the alarm to return to normal after it has been in alarm status.

_____ 6. The most common operator interface device that receives alarm notification is a ___.

_____ 7. Data trending is the use of past building equipment performance data to determine ___ system needs.

     A. present
     B. future
     C. yearly
     D. automated

_____ 8. A time interval of ___ min is commonly used for long-term data trending.

     A. 10
     B. 20
     C. 30
     D. 60

_____ 9. ___ is scheduled inspection and work (lubrication, adjustment, cleaning) required to maintain equipment in peak operating condition.

_____ 10. Preventive maintenance is usually less expensive than the ___ approach.

T   F   11. Building automation system software can create preventive maintenance work orders associated with a specific input or output value.

_____ **12.** Building automation systems use building system management software which ___ building and equipment conditions to maintenance technicians.

_____ **13.** Some building management systems use ___ of a particular piece of equipment so that actual temperature, humidity, pressure, or equipment status values can be superimposed.

_____ **14.** ___ maintenance is the monitoring of wear conditions and equipment characteristics and comparing them to a predetermined tolerance to predict possible malfunctions or failures.

      A. Routine
      B. Computer
      C. Preventive
      D. Predictive

_____ **15.** ___ can be set up to monitor most inputs and outputs of a building automation system.

T   F   **16.** Data trends can be imported into a Microsoft® Word® document and used to create graphs and charts.

_____ **17.** The building system management documentation function is commonly used to record ___.

Building System Management Software

_____ **1.** incremental bar indicator

_____ **2.** button indicates equipment status

_____ **3.** button enables linking to other graphics/functions

_____ **4.** real-time value for live system

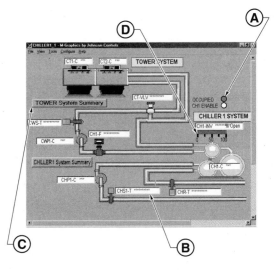

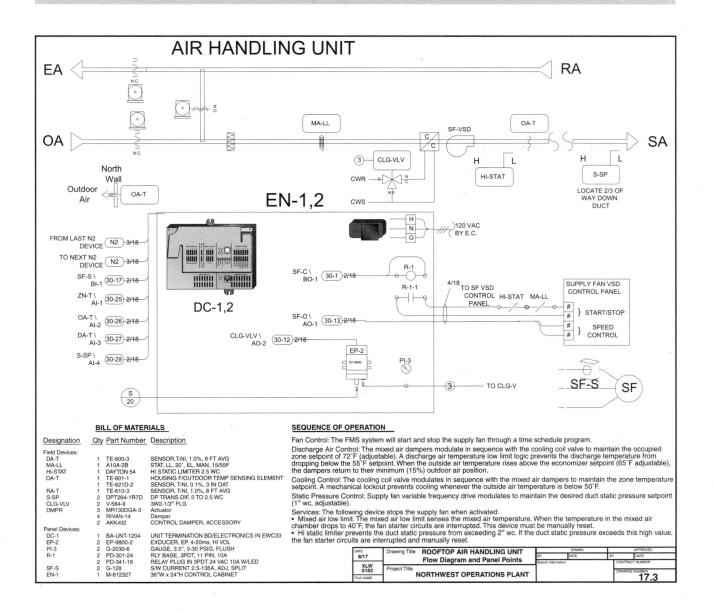

Activity 23-1. Alarm Value Setup

Use the rooftop air handling unit drawing to answer the questions.

BILL OF MATERIALS

Designation	Qty	Part Number	Description
Field Devices:			
DA-T	1	TE-600-3	SENSOR, T-NI, 1.0%, 8 FT AVG
MA-LL	1	A10A-2B	STAT, LL, 20", EL, MAN, 15/55F
HI-STAT	1	DAYTON 54	HI STATIC LIMITER 2.5 WC
OA-T	1	TE-601-1	HOUSING F/OUTDOOR TEMP SENSING ELEMENT
	1	TE-621D-2	SENSOR, T-NI, 0.1%, 3 IN OAT
RA-T	1	TE-610-3	SENSOR, T-NI, 1.0%, 8 FT AVG
S-SP	2	DPT264-1R7D	DP TRANS DIF, 0 TO 2.5 WC
CLG-VLV	2	V-584-9	3W2-1/2" FLG
DMPR	3	MR130DGA-3	Actuator
	4	RIVAN-14	Damper
	2	AKK402	CONTROL DAMPER, ACCESSORY
Panel Devices:			
DC-1	1	BA-UNT-1204	UNIT TERMINATION BD/ELECTRONICS IN EWC33
EP-2	2	EP-9800-2	EXDUCER, EP, 4-20ma, HI VOL
PI-3	2	G-2030-6	GAUGE, 3.5", 0-30 PSIG, FLUSH
R-1	2	PD-301-24	RLY BASE, 3PDT, 11 PIN, 10A
	2	PD-341-16	RELAY PLUG IN 3PDT 24 VAC 10A W/LED
SF-S	2	G-128	S/W CURRENT 2.5-135A, ADJ, SPLIT
EN-1	1	M-812327	36"W x 24"H CONTROL CABINET

SEQUENCE OF OPERATION

Fan Control: The FMS system will start and stop the supply fan through a time schedule program.

Discharge Air Control: The mixed air dampers modulate in sequence with the cooling coil valve to maintain the occupied zone setpoint of 72˚F (adjustable). A discharge air temperature low limit logic prevents the discharge temperature from dropping below the 55˚F setpoint. When the outside air temperature rises above the economizer setpoint (65˚F adjustable), the dampers return to their minimum (15%) outdoor air position.

Cooling Control: The cooling coil valve modulates in sequence with the mixed air dampers to maintain the zone temperature setpoint. A mechanical lockout prevents cooling whenever the outside air temperature is below 50˚F.

Static Pressure Control: Supply fan variable frequency drive modulates to maintain the desired duct static pressure setpoint (1" wc, adjustable).

Services: The following device stops the supply fan when activated:
• Mixed air low limit. The mixed air low limit senses the mixed air temperature. When the temperature in the mixed air chamber drops to 40˚F, the fan starter circuits are interrupted. This device must be manually reset.
• Hi static limiter prevents the duct static pressure from exceeding 2" wc. If the duct static pressure exceeds this high value, the fan starter circuits are interrupted and manually reset.

DATE 8/17	Drawing Title **ROOFTOP AIR HANDLING UNIT Flow Diagram and Panel Points**	DRAWN		APPROVED	
		BY	DATE	BY	DATE
XLW 0183		Branch Information		CONTRACT NUMBER	
FILE NAME	Project Title **NORTHWEST OPERATIONS PLANT**			DRAWING NUMBER **17.3**	

1. List the analog inputs for the air handling unit.

_____

_____

2. _____ Is a zone temperature occupied setpoint given in the sequence of operation?

3. _____ The zone temperature occupied setpoint is ___°F.

4. _____ The zone temperature occupied alarm low limit is ___°F.

5. _____ The zone temperature occupied alarm high limit is ___°F.

6. _____ The zone temperature occupied alarm differential is ___°F.

7. _____ The zone temperature occupied alarm delay is ___ min.

8. _____ Is a duct static pressure setpoint given in the sequence of operation?

9. _____ The duct static pressure setpoint is ___″ wc.

10. _____ The duct static pressure alarm low limit is ___″ wc.

11. _____ The duct static pressure alarm high limit is ___″ wc.

12. _____ The duct static pressure alarm differential is ___″ wc.

13. _____ The duct static pressure setpoint delay is ___ min.

14. _____ Is a discharge air temperature setpoint given in the sequence of operation?

15. _____ The discharge air temperature setpoint is ___°F.

16. _____ The discharge air temperature alarm low limit is ___°F.

17. _____ The discharge air temperature alarm high limit is ___°F.

18. _____ The discharge air temperature alarm differential is ___°F.

19. _____ The discharge air temperature setpoint delay alarm is ___ min.

20. _____ What type of point is the fan status switch?

21. _____ Would an alarm be created for it?

22. _____ What position would the switch be in when in alarm?

Activity 23-2. Alarm Message Setup

Use the critical alarm notification to answer the questions.

1. Write an appropriate alarm message for the zone temperature.

    _____

2. Write an appropriate alarm message for the duct static pressure.

    _____

    _____

CRITICAL ALARM	02/28	▭ ▱ ✕
CRITICAL ALARM RECEIVED		14:03:46

HI ALARM	AHU1\ZN-TEMP:ZONE TEMP	122.5 DEGF

Look Later	Acknowledge	Discard Alarm	Look Now	View System

3. Write an appropriate alarm message for the discharge air temperature.

    _____

    _____

Activity 23-3. Using Data Trending

Use the air handling unit temperature data trends to answer the questions.

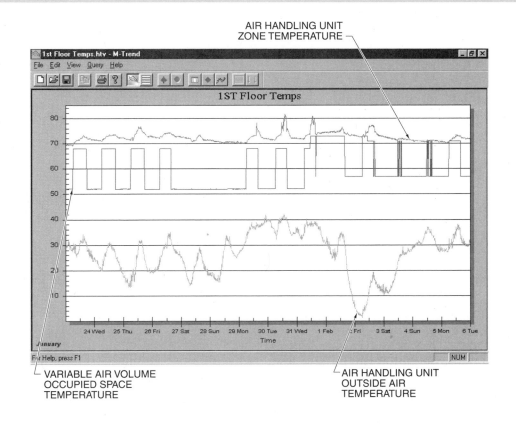

AIR HANDLING UNIT
ZONE TEMPERATURE

VARIABLE AIR VOLUME
OCCUPIED SPACE
TEMPERATURE

AIR HANDLING UNIT
OUTSIDE AIR
TEMPERATURE

1. _____ At what time did the highest air handling unit zone temperature occur?

2. _____ The zone temperature at that time was ___°F.

3. _____ The outside air temperature at that time was ___°F.

4. _____ What relationship exists between the zone temperature and outside air temperature?

5. List additional air handling unit information that may be gathered using data trends.

Activity 23-4. Preventive Maintenance Work Order Setup

Use the preventive maintenance work order and checklist to answer the questions.

PREVENTIVE MAINTENANCE WORK ORDER		
Work Order No: **46**	Requisition No:	
Issue Date: **6\21**　　Time: **15:49**	Skill: **Mechanical**	
Equipment Name: **Air Unit #3**	EQUIPMENT LOCATION	
Manufacturer: **Trane**	Building: **3E**	Floor: **3rd**
Description: **Air Handling Unit, 4 HP Century Motor**	System:	
Model No: **V4517**　Serial No: **100-AHU01**	Date Work Performed:	

LUBRICATION	COMMENTS	
Grease motor bearings		☐
Grease fan bearings		☐
Oil damper pivots		☐
Oil damper pneumatics		☐

PREVENTIVE MAINTENANCE CHECKLIST		
MECHANICAL	COMMENTS	
Lockout/Tagout power supply		☐
Inspect motor rotation (bearings)		☐
Inspect fan rotation (bearings)		☐
Inspect fan blades		☐
Inspect V-belts		☐
Inspect motor and fan sheaves		☐
Inspect dampers		☐
Inspect damper actuators		☐
Replace unit filters		☐
Vacuum debris from unit		☐

1. List the tools required to perform the indicated preventive maintenance on the air handling unit.

_____

_____

2. List items consumed during preventive maintenance on an air handling unit.

_____

_____

3. List a minimum of three safety considerations when performing preventive maintenance on an air handling unit.

_____

_____

Name: _____ Date: _____

Control System Principles

_____ 1. The parts of a commercial electric utility bill are ___.

    A. electrical consumption
    B. electrical demand
    C. power factor/fuel recovery
    D. all of the above

_____ 2. ___ is the total amount of electricity used during a billing period.

_____ 3. ___ is the highest amount of electricity used during a specific period of time.

_____ 4. Electrical consumption is measured in ___.

    A. kilovolts (kV)
    B. amps (A)
    C. kilowatt-hours (kWh)
    D. gigavolt hours (gVh)

T   F 5. Integrated demand is electrical demand calculated as an average over the time interval.

_____ 6. Power factor is a measure of electrical efficiency and is commonly expressed as a ___.

_____ 7. ___ is the amount of money a utility is permitted to charge to reflect the constantly changing cost of energy.

_____ 8. A(n) ___ is the classification of a customer depending on the type of service and the amount of electricity used.

    A. tier
    B. ratchet clause
    C. energy cost index
    D. exclusion

_____ 9. A(n) ___ is documentation that permits an electric utility to charge for demand based on the highest amount of electricity used in a 12-month period, not the amount actually measured.

_____ 10. A(n) ___ rate structure may be used to increase or decrease the cost of power depending on the time of day in which the electricity is used.

_____ **11.** ___ is a colorless, odorless fossil fuel.

    A. Gasoline
    B. Fuel oil
    C. Natural gas
    D. none of the above

_____ **12.** A therm is the quantity of gas required to produce ___ Btu.

    A. 100
    B. 1000
    C. 10,000
    D. 100,000

T   F   **13.** The energy cost index (ECI) is the amount of heat energy (in Btu) used in a commercial building divided by the number of square feet in the building.

_____ **14.** A(n) ___ is the amount of energy required to raise the temperature of 1 lb of water 1°F.

_____ **15.** A(n) ___ is an inventory of the energy-consuming equipment in a commercial building.

T   F   **16.** Observing the dress of individuals in a building can indicate current building condition comfort levels.

_____ **17.** Building automation system proposals are written to ___ HVAC equipment and services such as furnaces, water treatment equipment, and preventive maintenance.

_____ **18.** A(n) ___ is a one-page summary of the significant parts of a proposal.

    A. exclusion
    B. executive summary
    C. introduction
    D. signature sheet

_____ **19.** A(n) ___ is an item in a proposal that is the responsibility of the customer and not the contractor.

_____ **20.** A critical part of an executive summary is the ___.

Natural Gas Bill

_____ **1.** natural gas cost

_____ **2.** environmental recovery cost

_____ **3.** customer charge

_____ **4.** distribution (delivery) charge

Ⓓ   Ⓐ

ACME Utility Company
Bill for Natural Gas Service
Customer Name:
Jackson Memorial Hospital
123 Main Street

Meter ID Number: 123456789
Commercial

Bill Date: 7/24

Natural Gas Usage: 40 cu ft
Conversion to Therms:
40 × 1.010 Btu Factor =
40.40 Therms

Current Charges
Monthly Customer Charge   54.50
Distribution Charge: First 40 Therms   40.40 @ .1330 = 5.37
Environmental Recovery Cost   40.40 @ .0046 = .19
                                               $60.06

Natural Gas Cost
40.40 Therms × .3961

                                              $16.00
Taxes
State Tax   40.40 @ .024 = .97
Utility Fund Tax   76.06 @ .001 = .08
                                               $1.05

Total current bill-est   **$77.11**

Ⓒ   Ⓑ

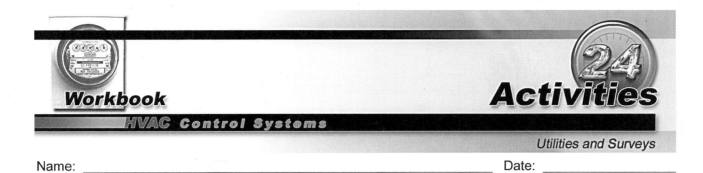

Workbook

HVAC Control Systems

24 Activities

Utilities and Surveys

Name: _____ Date: _____

Activity 24-1. Electrical Cost Determination

It is summer and a customer has called indicating that their utility bill is high. The customer wants a breakdown of their utility bill. Use the meter readings and the monthly utility rates to calculate the electrical cost.

MONTHLY UTILITY RATES

KILOWATT DEMAND CHARGE (IN DOLLARS PER kW)		
	SUMMER	WINTER
FOR FIRST 50 kW	$9.62	$8.62
FOR ALL EXCESS OVER 50 kW	$8.34	$7.43

KILOWATT HOUR CHARGE (IN CENTS PER kWh)		
	SUMMER	WINTER
FOR FIRST 40,000 kWh	4.58	4.15
FOR NEXT 60,000 kWh	3.27	2.95
FOR NEXT 200 kWh PER kWd BUT NOT LESS THAN 400,000 kWh	2.86	2.66
FOR NEXT 200 kWh PER kWd	2.182	2.043
FOR ALL EXCESS	.91	.82

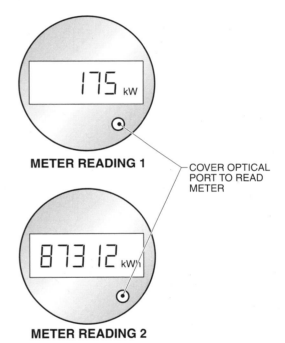

METER READING 1

COVER OPTICAL PORT TO READ METER

METER READING 2

1. _____ The total usage cost for the month is $___.

2. _____ The total demand cost for the month is $___.

3. _____ If the fuel recovery charge is $ .02 per kWh, the fuel recovery charge for the month is ___.

4. _____ The total electric bill for the month, excluding other taxes and power factor charges, is ___.

5. _____ If the building is 10,000 sq ft, the ECI for the month is $___ per sq ft.

Activity 24-2. Energy Use Analysis

A company has provided last year's utility bills. Plot the company energy use data to help justify expenditures on new mechanical equipment.

MONTHLY ELECTRICAL USAGE/DEMAND		
MONTH	USAGE*	DEMAND†
January	65,500	140
February	67,000	145
March	72,000	155
April	75,000	165
May	80,000	171
June	83,000	177
July	87,000	175
August	89,000	180
September	84,000	165
October	84,000	163
November	72,000	154
December	68,000	152

* in kWh

† in kW

1. Graph the company energy usage data.

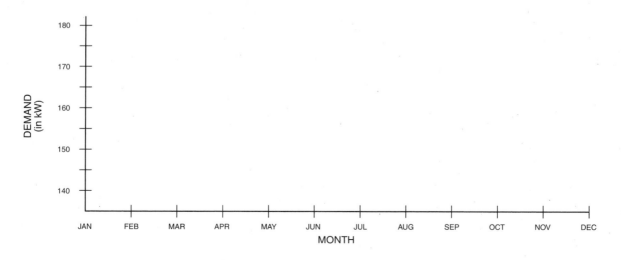

2. Graph the company energy demand data.

Name: _____ Date: _____

Control System Principles

_____ **1.** A building automation system ___ should be consulted regarding specific procedures for troubleshooting equipment.

_____ **2.** Manufacturers provide an override indicator such as a(n) ___ in the building automation system control software that can be accessed from the workstation computer that indicates an override is in effect.

    A. XX
    B. II
    C. O
    D. 1

_____ **3.** The primary cause of excessively hot or cold temperature complaints is having the ___ in the building automation system controller software.

T   F **4.** The standard static pressure setpoint for a variable air volume system is 5″ wc.

_____ **5.** When using a variable air volume system, the discharge air temperature should be checked at the air handling unit to ensure it is at ___°F.

    A. 55
    B. 65
    C. 74
    D. 76

_____ **6.** To protect sensors, manufacturers sell ___ that protect a sensor from damage.

_____ **7.** Improper or no sensor reading may be caused by ___ in the control software.

T   F **8.** Manufacturers provide software that is able to tune control loops to provide the proper control and eliminate constant cycling of controlled devices.

_____ **9.** When a mechanical system is providing inlet water that is rapidly fluctuating in temperature, the ___ cycle rapidly in response.

_____ **10.** Too many or too few alarms may be caused by a failure to ___ alarms properly when setting up the system.

Building Automation System Software Indicators

_____ **1.** operating indicator

_____ **2.** override indicator

_____ **3.** shutdown indicator

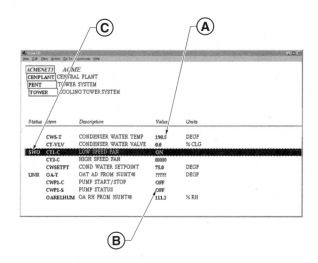

Alarm Parameter Folders

_____ **1.** parameter

_____ **2.** setpoint

_____ **3.** technician name code

_____ **4.** alarmed sensor

Communication Bus Troubleshooting

_____ **1.** network connection

_____ **2.** network communication module

_____ **3.** controller

_____ **4.** local communication bus

_____ **5.** technician workstation

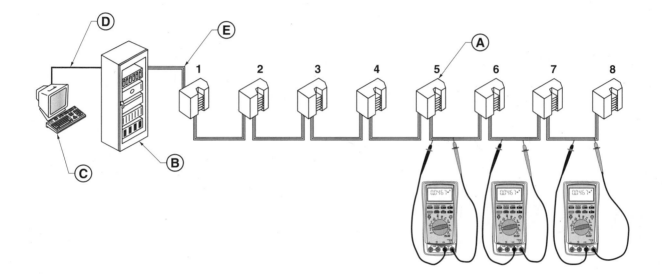

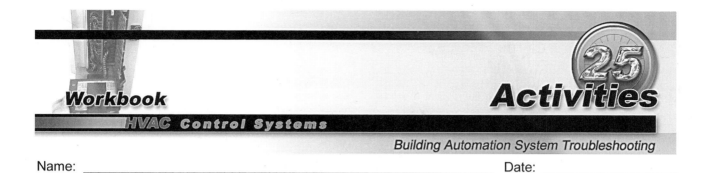

Name: _____ Date: _____

Activity 25-1. Sensor Troubleshooting

While checking a building automation system operator workstation screen, a system sensor appears to have a problem. Use the operator workstation display to answer the questions.

Status	Item	Description	Value	Units
	CWS-T	CONDENSER WATER TEMP	190.5	DEGF
	CT-VLV	CONDENSER WATER VALVE	0.0	% CLG
SWO	CT1-C	LOW SPEED FAN	ON	
	CT2-C	HIGH SPEED FAN	00000	
	CWSETPT	COND WATER SETPOINT	75.0	DEGF
UNR	OA-T	OAT AD FROM 35UNT40	?????	DEGF
	CWP1-C	PUMP START/STOP	OFF	
	CWP1-S	PUMP STATUS	OFF	
	OARELHUM	OA RH FROM 35UNT40	111.2	% RH

OPERATOR WORKSTATION DISPLAY

SUNSHIELD

TEMPERATURE SENSOR SENSING ELEMENT

TO CONTROLLER {

1. _____ The ___ sensor (item) has question marks in the value column.

2. _____ The three letters that are listed in the status column of the sensor are ___.

3. _____ What might this mean?

4. What is the next logical step since it is suspected that this sensor has a problem?

_____

_____

A DMM is used to test the sensor after determining its location in the system. The sensor is listed as a PTC sensor that has a nominal value of 1000 Ω at 70°F and a coefficient of 2.2 Ω per °F. The outside air temperature is 85°F.

5. The sensor resistance should be ___ Ω.

> *The sensor is disconnected from the building automation system controller, and a DMM set to measure resistance is used to check the sensor. The sensor resistance is 3 Ω.*

6. _____ What is the problem?

7. _____ What should be done?

Activity 25-2. Building Automation System Software Troubleshooting

> *While checking a building automation system operator workstation screen, it is determined that a tower low-speed fan is ON when it should be OFF. Use the operator workstation display to answer the questions.*

1. _____ The tower current value listed on the display is ___.

2. _____ The tower listed command priority is ___.

3. _____ The tower commanded feature is ___.

4. _____ The tower communication status is ___.

5. _____ The listed S/W override (in the software) is ___.

6. _____ Does this indicate that the fan has been overridden by an operator?

7. What should be done to correct the problem?

OPERATOR WORKSTATION DISPLAY

Activity 25-3. Troubleshooting a Constantly Cycling Control Device

A maintenance technician has noticed that a heating valve is constantly cycling open and closed. A laptop computer is used to open the software folder for the controller. Use the controller software folder information to answer the questions.

1. _____ The controller Occ Htg Setpt is ___.

2. _____ The controller Htg Prop Band is ___.

3. _____ The controller Htg Integ Time is ___.

CONTROLLER SOFTWARE FOLDER

The original controller software that was used when the heating loop worked properly was found. The original values were 68°F for Occ Htg Setpt, −10°F for Htg Prop Band, and 20 for Htg Integ Time.

4. What should be done?

_____

_____

5. What should be done to ensure proper operation?

_____

_____

Name: _____ Date: _____

Control System Principles

T   F   **1.** Poor system control is the result of insufficient time spent fine-tuning a system.

_____ **2.** Advanced HVAC control technologies help reduce control system ___ and improve control system accuracy.

_____ **3.** ___ is the ability of a controller to self-diagnose and self-correct a system in order to correct various problems.

      A. Fuzzy logic
      B. Web-enabled
      C. Network layering
      D. Echelon

_____ **4.** Self-tuning controllers release ___ from tuning control loops.

_____ **5.** ___ is the use of advanced computing power to make decisions about system setup and functionality.

_____ **6.** ___ communication is the use of a high-frequency electronic signal to communicate between control system components.

      A. Infrared
      B. Intranet
      C. PC-based
      D. Radio frequency

_____ **7.** A(n) ___ standard is published by an established governing body and adopted by a national or international standards group.

      A. controller
      B. closed system
      C. open protocol
      D. intranet

_____ **8.** The ___ layer of the open systems integration (OSI) seven-layer model defines the layout of devices such as hubs, repeaters, and wiring.

_____ **9.** ___ developed and maintains the Building Automation and Control Networks (BACnet).

T   F   **10.** A computer gateway is a single computer loaded with special software that permits the viewing of different manufacturers' software.

_____ **11.** ___ is the ability of devices produced by different manufacturers to communicate and share information.

_____ **12.** A ___ is an analog input, analog output, digital (binary) input, or digital output, each of which has different characteristics.

    A. map
    B. point
    C. protocol
    D. network

_____ **13.** BACnet has approximately ___ standard object classes.

_____ **14.** The ___ object contains sensor input.

    A. trend log
    B. notification class
    C. analog output
    D. analog input

_____ **15.** A protocol implementation ___ statement is a detailed description for a specific manufacturer's BACnet device that states its BACnet capabilities.

_____ **16.** The ___ is a protocol for obtaining interoperability in building automation systems using software and hardware.

T   F   **17.** Infrared communication is communication between devices using infrared (IR) radiation.

_____ **18.** A ___ is a small, hand-held computing device with a small display screen.

    A. personal digital assistant (PDA)
    B. personal computer (PC)
    C. router
    D. hub

_____ **19.** Artificial intelligence is used to reduce the number of ___systems that are installed in buildings.

_____ **20.** A(n) ___ contains a whole system-on-a-chip with multiple processors, read-write and read-only memory, communications, and input/output subsystems.

_____ **21.** A category of a BACnet interoperability building block (BIBB) is ___.

    A. data sharing
    B. scheduling
    C. trending
    D. all of the above

T   F   **22.** A network variable is a data item that a device application program will receive from or send to other devices on a network.

_____ **23.** LON devices are separated into groups according to their ___.

_____ **24.** In the building automation system installation process, a ___ identifies the field devices and field points to a head-end supervisory device.

      A. point map
      B. field controller
      C. network variable
      D. none of the above

T   F   **25.** One disadvantage of LON is that is not an ANSI standard.

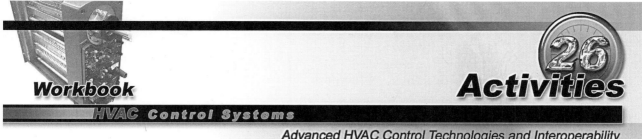

Name: _____   Date: _____

PROTOCOL IMPLEMENTATION CONFORMANCE STATEMENT

Table A: BACnet Protocol Implementation Conformance Statement

Vendor Name:	ABC Controls
Product Name:	NC35
Model Numbers:	NC35-0, NC35-1

Table B: BACnet Conformance Class Supported

Class 1	☐	Class 4	■
Class 2	☐	Class 5	☐
Class 3	☐	Class 6	☐

Table C: BACnet Functional Groups Supported

Clock	■	Files	☐
HHWS	☐	Reinitialize	■
PCWS	☐	Virtual Operator Interface	☐
Event Initiation	■	Virtual Terminal	☐
Event Response	■	Device Communications	☐
COV Event Initiation	☐	Time Master	☐
COV Event Response	☐		

Product Description

The ABC NC35 supervisory controller is designed to manage a small building or campus of buildings. The NC35 efficiently supervises the networking of application specific controllers (ASCs) and provides facility management features including weekly scheduling, alarm management, optimal start, and trending.

Facility personnel can review the system status and modify control parameters for the NC35 supervisory controller and its associated ASCs using a VT100 terminal or a graphical workstation.

With the addition of a network card, multiple NC35's can communicate over an Ethernet network, providing increased functionality for complex systems.

Table D: BACnet Standard Application Services Supported

Application Service	Initiates Requests	Executes Requests	Application Service	Initiates Requests	Executes Requests
Acknowledge Alarm	■	■	Read Range Service	☐	☐
Confirmed COV Notification	☐	☐	Write Property	■	■
Confirmed Event Notification	■	■	Write Property Multiple	■	■
Get Alarm Summary	☐	■	Device Communication Control	☐	☐
Get Enrollment Summary	☐	■	Confirmed Private Transfer	■	■
Get Event Information	☐	■	Unconfirmed Private Transfer	■	■
Subscribe COV	☐	☐	Reinitialize Device	☐	■
Subscribe COV Property	☐	☐	Confirmed Text Message	☐	☐
Unconfirmed COV Notification	■	■	Unconfirmed Text Message	☐	☐
Unconfirmed Event Notification	■	■	UTC Time Synchronization	■	■
			Time Synchronization	☐	☐
Atomic Read File	☐	☐	Who-Has	■	■
Atomic Write File	☐	☐	I-Have	■	■
			Who-Is	■	■
Add List Element	■	■	I-Am	■	■
Remove List Element	■	■	VT-Open	☐	☐
Create Object	■	■	VT-Close	☐	☐
Delete Object	■	■	VT-Data	☐	☐
Read Property	■	■			
Read Property Conditional	☐	☐	Authenticate	☐	☐
Read Property Multiple	■	■	Request Key	☐	☐

Activity 26-1. BACnet Protocol

A hospital is considering the replacement of an old DDC system. An information sheet (PICS document) has been provided describing the BACnet capabilities of a possible new controller. Use the PICS document to answer the questions.

1. _____ The controller manufacturer is ___.

2. _____ The controller (product) name is ___.

3. _____ Can a VT100 terminal be used as an operator interface?

4. _____ The BACnet conformance class that is supported is ___.

5. List the BACnet functional groups that are supported.

_____

_____

6. _____ Does the controller support authenticate?

7. _____ Does the controller support acknowledge alarm?

8. _____ Does the controller support who-has?

9. _____ Does the controller support VT-open?

10. _____ Does the controller support read property?

11. _____ Does the controller support subscribe COV?

12. _____ Does the controller support write property?

13. _____ Does the controller support read range service?

14. _____ Is the PICS document valid for other controllers manufactured by ABC Controls?

15. _____ Could future upgrades of the NC35 controller add more features?

16. _____ A facility wishes to integrate a chiller control system that is listed as BACnet compatible. Does this automatically mean that all of the objects and features can be shared between the NC35 and the chiller controller?

17. _____ What document should be obtained from the chiller controller manufacturer?

18. _____ A facility specification includes the phrase, "Controllers must be BACnet compatible." Is this wording adequate?